KUMON

Grades 2-4

Math Boosters

Multiplication & Division

W9-AZS-023

To parents

This workbook is designed for children to complete by themselves. By checking their answers and correcting errors on their own, children can strengthen their independence and develop into self-motivated learners.

At Kumon, we believe that each child should do work according to his or her ability, rather than his or her age or grade level. So, if this workbook is too difficult or too easy for your child, please choose another Kumon Math Workbook with an appropriate level of difficulty.

How to use this book

1. **Let's get started!** Start by writing the date on the top of each page so you can track your progress.
2. **Let's go!** Start with Step 1 and review what you already know.
3. **Don't forget!** Read the "Don't Forget" boxes which contain helpful explanations and examples.
4. **Let's work!** Solve the problems in numerical order step-by-step. Look at the samples problems for help and return to the "Review" box if you need to refresh your knowledge.
5. **Let's check your answers!** After you have finished solving the problems, check your answers and add up your score on each page. If you don't know how to do this, ask your parent or guardian to show you.
6. **Let's get it right!** Once you have finished checking your answers, review any errors to see where you made a mistake and then try again.

Math Boosters   Grades 2-4 Multiplication & Division

● Table of Contents ●

1	Multiplication 2× to 9×	2×, 3×	4
2		4×, 5×	6
3		6×, 7×	8
4		8×, 9×	10
TEST		Multiplication 2× to 9×	12
5	Multiplication ×1-Digit	2-Digits ×1-Digit 1	14
6		2-Digits ×1-Digit 2	16
7		2-Digits ×1-Digit 3	18
8		2-Digits ×1-Digit 4	20
9		2-Digits ×1-Digit 5	22
10		2-Digits ×1-Digit 6	24
11		2-Digits ×1-Digit 7	26
12		2-Digits ×1-Digit 8	28
TEST		Multiplication ×1-Digit (2-Digits×1-Digit)	30
13		3-Digits ×1-Digit 1	32
14		3-Digits ×1-Digit 2	34
15		3-Digits ×1-Digit 3	36
16		4-Digits ×1-Digit	38
TEST		Multiplication ×1-Digit (3-Digits×1-Digit, 4-Digits×1-Digit)	40
17	Multiplication ×2-Digits	2-Digits ×2-Digits 1	42
18		2-Digits ×2-Digits 2	44
19		2-Digits ×2-Digits 3	46
20		2-Digits ×2-Digits 4	48
21		2-Digits ×2-Digits 5	50
22		2-Digits ×2-Digits 6	52
23		2-Digits ×2-Digits 7	54
24		3-Digits ×2-Digits 1	56
25		3-Digits ×2-Digits 2	58
26		3-Digits ×2-Digits 3	60
TEST		Multiplication ×2-Digits	62
27	Mental Math Division	Inverse Multiplication 1	64
28		Inverse Multiplication 2	66
29		Inverse Multiplication 3	68

30		Division with Remainders 1	70
31		Division with Remainders 2	72
32		Division with Remainders 3	74
TEST		Mental Math Division	76
33	Division ÷1-Digit	2-Digits ÷1-Digit 1	78
34		2-Digits ÷1-Digit 2	80
35		2-Digits ÷1-Digit 3	82
36		2-Digits ÷1-Digit 4	84
37		3-Digits ÷1-Digit 1	86
38		3-Digits ÷1-Digit 2	88
39		3-Digits ÷1-Digit 3	90
TEST		Division ÷1-Digit	92
40	Division 2-Digits ÷2-Digits	2-Digits ÷2-Digits 1	94
41		2-Digits ÷2-Digits 2	96
42		2-Digits ÷2-Digits 3	98
43		2-Digits ÷2-Digits 4	100
44		2-Digits ÷2-Digits 5	102
45		2-Digits ÷2-Digits 6	104
46		2-Digits ÷2-Digits 7	106
TEST		Division 2-Digits ÷2-Digits	108
47	Division 3-Digits÷2-Digits	3-Digits ÷2-Digits 1	110
48		3-Digits ÷2-Digits 2	112
49		3-Digits ÷2-Digits 3	114
50		3-Digits ÷2-Digits 4	116
51		3-Digits ÷2-Digits 5	118
52		3-Digits ÷2-Digits 6	120
53		÷3-Digits	122
54		4-Digits ÷2-Digits	124
TEST		Division 3-Digits÷2-Digits	126
Answer Key			128

Multiplication 2× to 9×

2×, 3×

Date 4/20/2014

Score /100

Fill in the missing numbers in the boxes below.

> Let's think about how many times each number increases one by one.

(1) 2 — 4 — 6 — 8 — 10 — 12

(2) 10 — 12 — 14 — 16 — 18 — 20

(3) 3 — 6 — 9 — 12 — 15 — 18

1 **Fill in the boxes while reading the number sentences below.**

2 points per question

(1) $2 \times 1 = 2$
Two times one is two.

(2) $2 \times 2 = 4$
Two times two is four.

(3) $2 \times 3 = 6$
Two times three is six.

(4) $2 \times 4 = 8$
Two times four is eight.

(5) $2 \times 5 = 10$
Two times five is ten.

(6) $2 \times 6 = 12$
Two times six is twelve.

(7) $2 \times 7 = 14$
Two times seven is fourteen.

(8) $2 \times 8 = 16$
Two times eight is sixteen.

(9) $2 \times 9 = 18$
Two times nine is eighteen.

(1) $3 \times 1 = 3$
Three times one is three.

(2) $3 \times 2 = 6$
Three times two is six.

(3) $3 \times 3 = 9$
Three times three is nine.

(4) $3 \times 4 = 12$
Three times four is twelve.

(5) $3 \times 5 = 15$
Three times five is fifteen.

(6) $3 \times 6 = 18$
Three times six is eighteen.

(7) $3 \times 7 = 21$
Three times seven is twenty-one.

(8) $3 \times 8 = 24$
Three times eight is twenty-four.

(9) $3 \times 9 = 27$
Three times nine is twenty-seven.

2 Calculate.

2 points per question

(1) $2 \times 1 = \boxed{2}$ (4) $2 \times 2 = \boxed{4}$ (7) $2 \times 7 = \boxed{14}$

(2) $2 \times 5 = \boxed{10}$ (5) $2 \times 4 = \boxed{8}$ (8) $2 \times 9 = \boxed{18}$

(3) $2 \times 3 = \boxed{6}$ (6) $2 \times 8 = \boxed{16}$

3 Calculate.

2 points per question

(1) $3 \times 1 = \boxed{3}$ (4) $3 \times 2 = \boxed{6}$ (7) $3 \times 5 = \boxed{15}$

(2) $3 \times 4 = \boxed{12}$ (5) $3 \times 7 = \boxed{21}$ (8) $3 \times 9 = \boxed{27}$

(3) $3 \times 6 = \boxed{18}$ (6) $3 \times 3 = \boxed{9}$ (9) $3 \times 8 = \boxed{24}$

4 Fill in the missing numbers in the boxes below.

3 points per question

(1) $2 \times \boxed{3} = 6$ (6) $2 \times \boxed{9} = 18$

(2) $2 \times \boxed{4} = 8$ (7) $2 \times \boxed{8} = 16$

(3) $2 \times \boxed{5} = 10$ (8) $3 \times \boxed{7} = 21$

(4) $3 \times \boxed{1} = 3$ (9) $3 \times \boxed{6} = 18$

(5) $3 \times \boxed{2} = 6$ (10) $3 \times \boxed{5} = 15$

© Kumon Publishing Co., Ltd.   5

4×, 5×

Fill in the missing numbers in the boxes below.

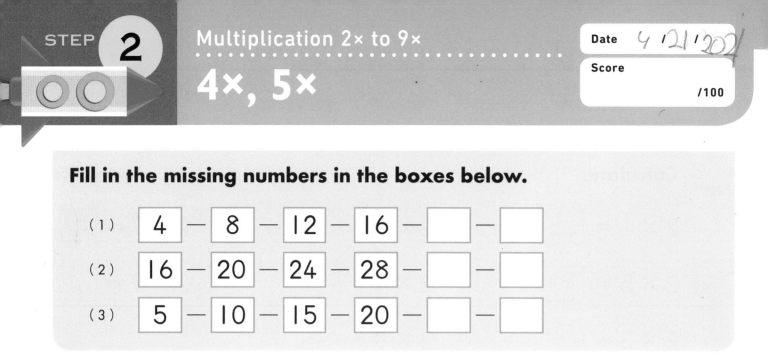

(1) 4 — 8 — 12 — 16 — [ ] — [ ]

(2) 16 — 20 — 24 — 28 — [ ] — [ ]

(3) 5 — 10 — 15 — 20 — [ ] — [ ]

1 **Fill in the boxes while reading the number sentences below.**

2 points per question

(1) $4 \times 1 =$ 4
Four times one is four.

(2) $4 \times 2 =$ 8
Four times two is eight.

(3) $4 \times 3 =$ 12
Four times three is twelve.

(4) $4 \times 4 =$ 16
Four times four is sixteen.

(5) $4 \times 5 =$ 20
Four times five is twenty.

(6) $4 \times 6 =$ 24
Four times six is twenty-four.

(7) $4 \times 7 =$ 28
Four times seven is twenty-eight.

(8) $4 \times 8 =$ 32
Four times eight is thirty-two.

(9) $4 \times 9 =$ 36
Four times nine is thirty-six.

(1) $5 \times 1 =$ 5
Five times one is five.

(2) $5 \times 2 =$ 10
Five times two is ten.

(3) $5 \times 3 =$ 15
Five times three is fifteen.

(4) $5 \times 4 =$ 20
Five times four is twenty.

(5) $5 \times 5 =$ 25
Five times five is twenty-five.

(6) $5 \times 6 =$ 30
Five times six is thirty.

(7) $5 \times 7 =$ 35
Five times seven is thirty-five.

(8) $5 \times 8 =$ 40
Five times eight is forty.

(9) $5 \times 9 =$ 45
Five times nine is forty-five.

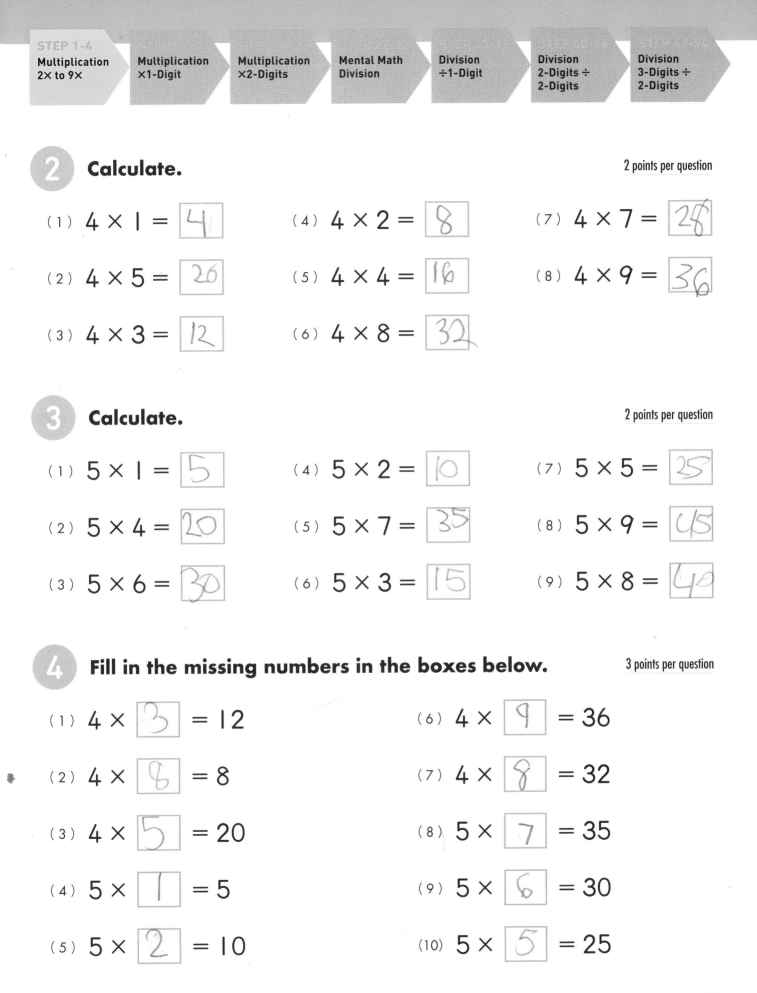

2 Calculate.

2 points per question

(1) $4 \times 1 = \boxed{4}$

(2) $4 \times 5 = \boxed{26}$

(3) $4 \times 3 = \boxed{12}$

(4) $4 \times 2 = \boxed{8}$

(5) $4 \times 4 = \boxed{16}$

(6) $4 \times 8 = \boxed{32}$

(7) $4 \times 7 = \boxed{28}$

(8) $4 \times 9 = \boxed{36}$

3 Calculate.

2 points per question

(1) $5 \times 1 = \boxed{5}$

(2) $5 \times 4 = \boxed{20}$

(3) $5 \times 6 = \boxed{30}$

(4) $5 \times 2 = \boxed{10}$

(5) $5 \times 7 = \boxed{35}$

(6) $5 \times 3 = \boxed{15}$

(7) $5 \times 5 = \boxed{25}$

(8) $5 \times 9 = \boxed{45}$

(9) $5 \times 8 = \boxed{40}$

4 Fill in the missing numbers in the boxes below.

3 points per question

(1) $4 \times \boxed{3} = 12$

(2) $4 \times \boxed{2} = 8$

(3) $4 \times \boxed{5} = 20$

(4) $5 \times \boxed{1} = 5$

(5) $5 \times \boxed{2} = 10$

(6) $4 \times \boxed{9} = 36$

(7) $4 \times \boxed{8} = 32$

(8) $5 \times \boxed{7} = 35$

(9) $5 \times \boxed{6} = 30$

(10) $5 \times \boxed{5} = 25$

6×, 7×

Fill in the missing numbers in the boxes below.

(1) 6 — 12 — 18 — 24 — 30 — 3[?]

(2) 30 — 36 — 42 — 48 — ☐ — ☐

(3) 7 — 14 — 21 — 28 — ☐ — ☐

1 Fill in the boxes while reading the number sentences below.

2 points per question

(1) 6 × 1 = 6
Six times one is six.

(2) 6 × 2 = 12
Six times two is twelve.

(3) 6 × 3 = 8
Six times three is eighteen.

(4) 6 × 4 = 24
Six times four is twenty-four.

(5) 6 × 5 = 30
Six times five is thirty.

(6) 6 × 6 = 36
Six times six is thirty-six.

(7) 6 × 7 = 42
Six times seven is forty-two.

(8) 6 × 8 = 48
Six times eight is forty-eight.

(9) 6 × 9 = 54
Six times nine is fifty-four.

(1) 7 × 1 = 7
Seven times one is seven.

(2) 7 × 2 = 14
Seven times two is fourteen.

(3) 7 × 3 = 21
Seven times three is twenty-one.

(4) 7 × 4 = 28
Seven times four is twenty-eight.

(5) 7 × 5 = 35
Seven times five is thirty-five.

(6) 7 × 6 = 42
Seven times six is forty-two.

(7) 7 × 7 = 49
Seven times seven is forty-nine.

(8) 7 × 8 = 56
Seven times eight is fifty-six.

(9) 7 × 9 = 63
Seven times nine is sixty-three.

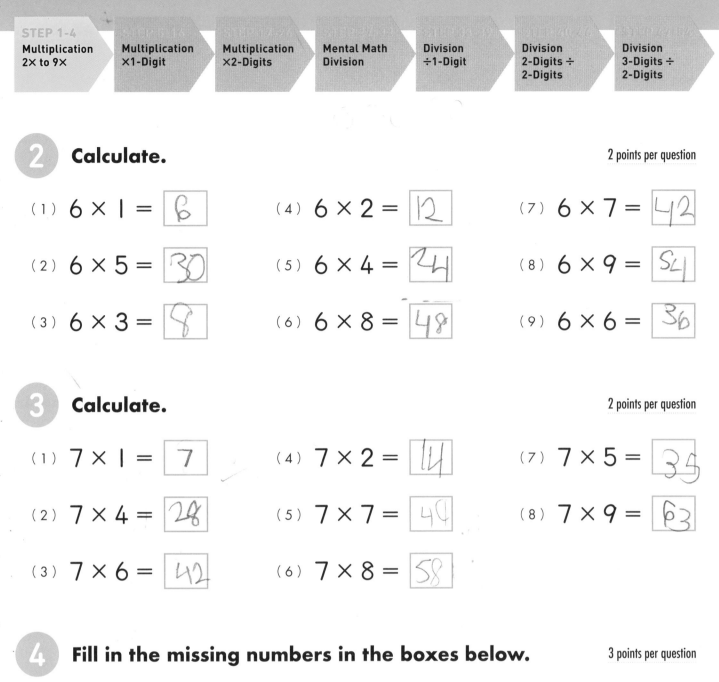

STEP 1-4
Multiplication 2× to 9×

Multiplication ×1-Digit

Multiplication ×2-Digits

Mental Math Division

Division ÷1-Digit

Division 2-Digits ÷ 2-Digits

Division 3-Digits ÷ 2-Digits

2 **Calculate.**

2 points per question

(1) $6 \times 1 = \boxed{6}$

(2) $6 \times 5 = \boxed{30}$

(3) $6 \times 3 = \boxed{8}$

(4) $6 \times 2 = \boxed{12}$

(5) $6 \times 4 = \boxed{24}$

(6) $6 \times 8 = \boxed{48}$

(7) $6 \times 7 = \boxed{42}$

(8) $6 \times 9 = \boxed{54}$

(9) $6 \times 6 = \boxed{36}$

3 **Calculate.**

2 points per question

(1) $7 \times 1 = \boxed{7}$

(2) $7 \times 4 = \boxed{28}$

(3) $7 \times 6 = \boxed{42}$

(4) $7 \times 2 = \boxed{14}$

(5) $7 \times 7 = \boxed{49}$

(6) $7 \times 8 = \boxed{58}$

(7) $7 \times 5 = \boxed{35}$

(8) $7 \times 9 = \boxed{63}$

4 **Fill in the missing numbers in the boxes below.**

3 points per question

(1) $7 \times \boxed{3} = 21$

(2) $7 \times \boxed{2} = 14$

(3) $7 \times \boxed{5} = 35$

(4) $6 \times \boxed{1} = 6$

(5) $6 \times \boxed{2} = 12$

(6) $7 \times \boxed{9} = 63$

(7) $7 \times \boxed{8} = 56$

(8) $6 \times \boxed{6} = 42$

(9) $6 \times \boxed{6} = 36$

(10) $6 \times \boxed{5} = 30$

8×, 9×

Date 5 / 14 / 20[?]

Score /100

Fill in the missing numbers in the boxes below.

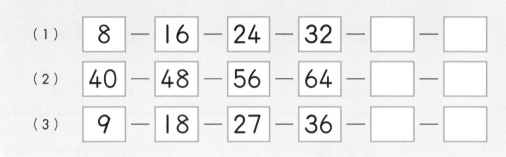

(1) 8 — 16 — 24 — 32 — ☐ — ☐

(2) 40 — 48 — 56 — 64 — ☐ — ☐

(3) 9 — 18 — 27 — 36 — ☐ — ☐

1 Fill in the boxes while reading the number sentences below.

2 points per question

(1) $8 \times 1 = \boxed{8}$
Eight times one is eight.

(2) $8 \times 2 = \boxed{16}$
Eight times two is sixteen.

(3) $8 \times 3 = \boxed{24}$
Eight times three is twenty-four.

(4) $8 \times 4 = \boxed{32}$
Eight times four is thirty-two.

(5) $8 \times 5 = \boxed{40}$
Eight times five is forty.

(6) $8 \times 6 = \boxed{45}$
Eight times six is forty-eight.

(7) $8 \times 7 = \boxed{56}$
Eight times seven is fifty-six.

(8) $8 \times 8 = \boxed{64}$
Eight times eight is sixty-four.

(9) $8 \times 9 = \boxed{}$
Eight times nine is seventy-two.

(1) $9 \times 1 = \boxed{9}$
Nine times one is nine.

(2) $9 \times 2 = \boxed{18}$
Nine times two is eighteen.

(3) $9 \times 3 = \boxed{27}$
Nine times three is twenty-seven.

(4) $9 \times 4 = \boxed{36}$
Nine times four is thirty-six.

(5) $9 \times 5 = \boxed{45}$
Nine times five is forty-five.

(6) $9 \times 6 = \boxed{44}$
Nine times six is fifty-four.

(7) $9 \times 7 = \boxed{63}$
Nine times seven is sixty-three.

(8) $9 \times 8 = \boxed{72}$
Nine times eight is seventy-two.

(9) $9 \times 9 = \boxed{8}$
Nine times nine is eighty-one.

STEP 1-4

| Multiplication 2× to 9× | Multiplication ×1-Digit | Multiplication ×2-Digits | Mental Math Division | Division ÷1-Digit | Division 2-Digits ÷ 2-Digits | Division 3-Digits ÷ 2-Digits |

2 Calculate.

2 points per question

(1) 8 × 1 = 8

(2) 8 × 5 = 40

(3) 8 × 3 = 24

(4) 8 × 2 = 16

(5) 8 × 4 = 32

(6) 8 × 8 = 27

(7) 8 × 7 = 56

(8) 8 × 9 = 72

(9) 8 × 6 = 48

3 Calculate.

2 points per question

(1) 9 × 1 = 9

(2) 9 × 4 = 36

(3) 9 × 6 = 84

(4) 9 × 2 = 10

(5) 9 × 7 = 32

(6) 9 × 3 = 64

(7) 9 × 5 = 56

(8) 9 × 9 = 72

4 Fill in the missing numbers in the boxes below.

3 points per question

(1) 8 × 3 = 24

(2) 8 × 2 = 16

(3) 8 × 5 = 40

(4) 9 × 1 = 9

(5) 9 × 2 = 18

(6) 8 × 9 = 72

(7) 8 × 8 = 64

(8) 9 × 7 = 63

(9) 9 × 6 = 54

(10) 9 × 5 = 45

Multiplication 2× to 9×

Date 0/6/8⁄

Score /100

Review STEP 1 STEP 2   **Calculate.**   2 points per question

(1) $2 \times 9 = 18$

(2) $3 \times 8 = 24$

(3) $4 \times 7 = 28$

(4) $5 \times 6 = 30$

(5) $2 \times 5 = 10$

(6) $3 \times 4 = 10$

(7) $4 \times 3 = 12$

(8) $5 \times 2 = 10$

(9) $3 \times 7 = 21$

(10) $5 \times 4 = 20$

(11) $3 \times 3 = 9$

(12) $5 \times 9 = 45$

(13) $2 \times 8 = 16$

(14) $4 \times 6 = 24$

(15) $5 \times 7 = 35$

(16) $4 \times 8 = 32$

(17) $3 \times 9 = 27$

(18) $2 \times 6 = 12$

(19) $4 \times 9 = 36$

(20) $3 \times 5 = 15$

(21) $2 \times 3 = 6$

(22) $4 \times 4 = 16$

(23) $2 \times 7 = 14$

(24) $5 \times 8 = 40$

(25) $3 \times 6 = 18$

Remember the two times table to the five times table.

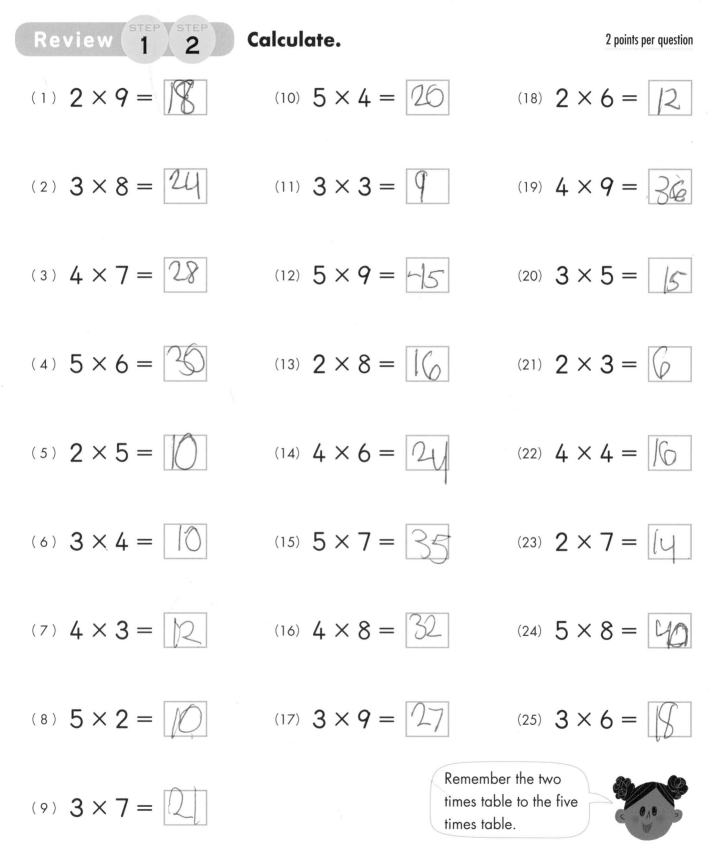

STEP 1-4

| Multiplication 2× to 9× | Multiplication ×1-Digit | Multiplication ×2-Digits | Mental Math Division | Division ÷1-Digit | Division 2-Digits ÷ 2-Digits | Division 3-Digits ÷ 2-Digits |

Review STEP 3 STEP 4   Calculate.

2 points per question

(1) $8 \times 7 = \boxed{56}$

(2) $7 \times 5 = \boxed{35}$

(3) $9 \times 7 = \boxed{63}$

(4) $6 \times 8 = \boxed{48}$

(5) $8 \times 6 = \boxed{48}$

(6) $6 \times 5 = \boxed{30}$

(7) $8 \times 2 = \boxed{16}$

(8) $6 \times 4 = \boxed{24}$

(9) $8 \times 9 = \boxed{72}$

(10) $7 \times 2 = \boxed{14}$

(11) $6 \times 7 = \boxed{42}$

(12) $8 \times 8 = \boxed{64}$

(13) $7 \times 7 = \boxed{49}$

(14) $7 \times 9 = \boxed{73}$

(15) $6 \times 9 = \boxed{54}$

(16) $8 \times 5 = \boxed{40}$

(17) $9 \times 2 = \boxed{18}$

(18) $9 \times 8 = \boxed{72}$

(19) $7 \times 6 = \boxed{42}$

(20) $9 \times 4 = \boxed{18}$

(21) $7 \times 3 = \boxed{21}$

(22) $8 \times 3 = \boxed{24}$

(23) $7 \times 8 = \boxed{56}$

(24) $9 \times 6 = \boxed{54}$

(25) $9 \times 9 = \boxed{81}$

If you got an answer wrong, let's go back to the table in the appropriate section and review.

STEP 5

Multiplication ×1-Digit

2-Digits×1-Digit 1

Date

Score

/100

Review STEP 1 STEP 2

Calculate.

(1) $3 \times 4 =$ ☐

(2) $2 \times 7 =$ ☐

(3) $4 \times 9 =$ ☐

(4) $2 \times 3 =$ ☐

(5) $4 \times 2 =$ ☐

(6) $4 \times 1 =$ ☐

1 Calculate.

6 points per question

Example • How to calculate 13×2.

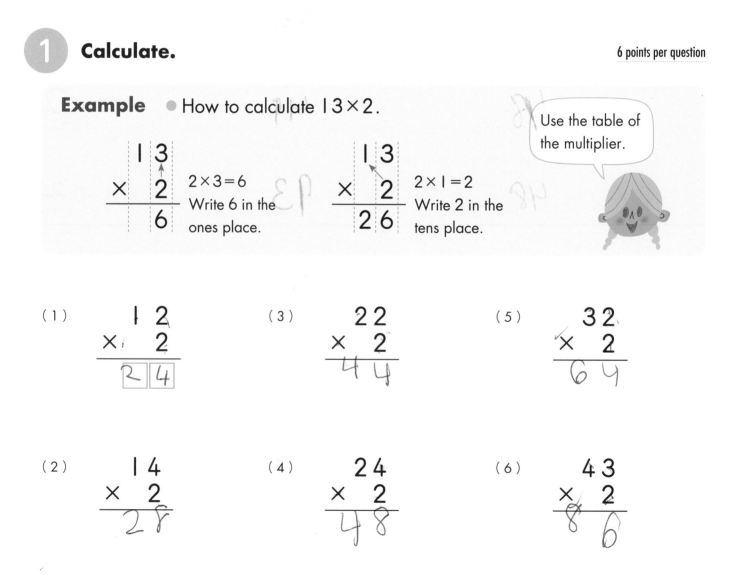

$$
\begin{array}{r}
1\,3 \\
\times\quad 2 \\
\hline
6
\end{array}
$$
$2 \times 3 = 6$
Write 6 in the ones place.

$$
\begin{array}{r}
1\,3 \\
\times\quad 2 \\
\hline
2\,6
\end{array}
$$
$2 \times 1 = 2$
Write 2 in the tens place.

Use the table of the multiplier.

(1)
$$
\begin{array}{r}
1\,2 \\
\times\quad 2 \\
\hline
2\,4
\end{array}
$$

(3)
$$
\begin{array}{r}
2\,2 \\
\times\quad 2 \\
\hline
4\,4
\end{array}
$$

(5)
$$
\begin{array}{r}
3\,2 \\
\times\quad 2 \\
\hline
6\,4
\end{array}
$$

(2)
$$
\begin{array}{r}
1\,4 \\
\times\quad 2 \\
\hline
2\,8
\end{array}
$$

(4)
$$
\begin{array}{r}
2\,4 \\
\times\quad 2 \\
\hline
4\,8
\end{array}
$$

(6)
$$
\begin{array}{r}
4\,3 \\
\times\quad 2 \\
\hline
8\,6
\end{array}
$$

STEP		STEP 5-16		STEP		STEP	
Multiplication 2× to 9×		Multiplication ×1-Digit		Multiplication ×2-Digits		Mental Math Division	

Division ÷1-Digit · Division 2-Digits ÷ 2-Digits · Division 3-Digits ÷ 2-Digits

2 Calculate.

6 points per question

(1)
```
   1 1
 ×   3
 ───
  3 3
```

(3)
```
   2 1
 ×   3
 ───
  6 3
```

(5)
```
   3 1
 ×   3
 ───
  9 3
```

(2)
```
   2 2
 ×   3
 ───
  6 6
```

(4)
```
   2 3
 ×   2
 ───
  4 6
```

(6)
```
   3 2
 ×   3
 ───
  9 6
```

3 Calculate.

7 points per question

(1)
```
   1 2
 ×   4
 ───
  4 8
```

(3)
```
   2 1
 ×   4
 ───
  8 4
```

(2)
```
   1 1
 ×   4
 ───
  4 4
```

(4)
```
   2 2
 ×   4
 ───
  8 8
```

Remember the four times table.

© Kumon Publishing Co., Ltd.   15

Review STEP 5

Calculate.

(1)
```
  1 2
×   2
```

(2)
```
  2 3
×   3
```

(3)
```
  1 1
×   4
```

---

1 Calculate.

5 points per question

Example ● How to calculate 16×2.

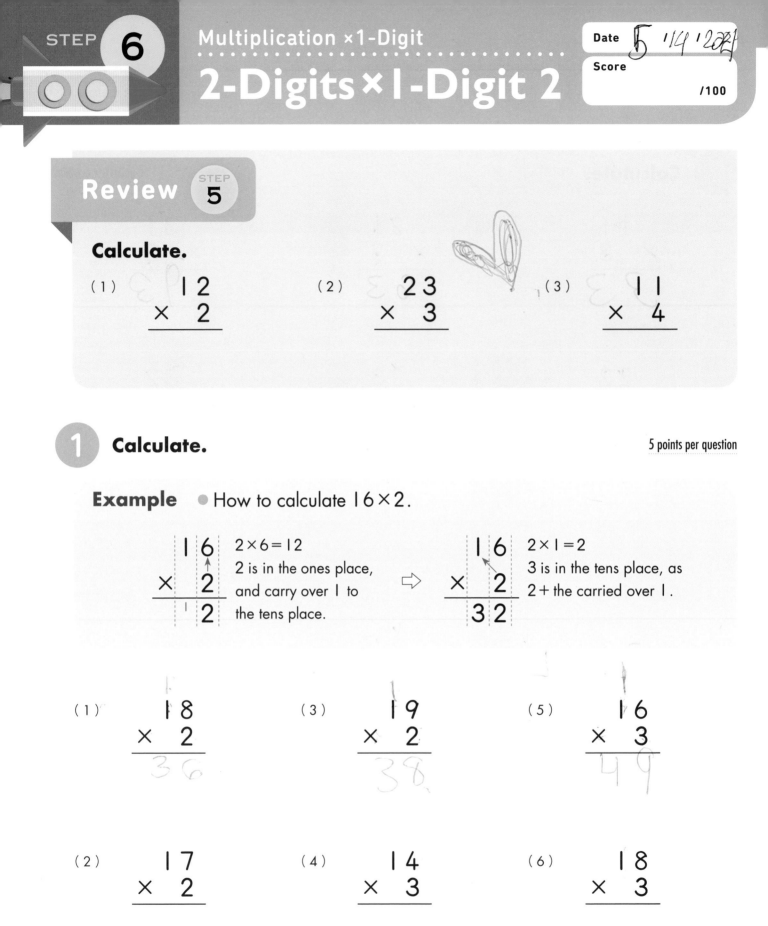

```
  1 6
×   2
──────
  1 2
```
2×6=12
2 is in the ones place, and carry over 1 to the tens place.

⇨

```
  1 6
×   2
──────
  3 2
```
2×1=2
3 is in the tens place, as 2+ the carried over 1.

(1)
```
  1 8
×   2
──────
  3 6
```

(3)
```
  1 9
×   2
──────
  3 8
```

(5)
```
  1 6
×   3
──────
  4 9
```

(2)
```
  1 7
×   2
```

(4)
```
  1 4
×   3
```

(6)
```
  1 8
×   3
```

2 Calculate.

5 points per question

(1)
$$\begin{array}{r} 26 \\ \times\ 2 \\ \hline \end{array}$$

(4)
$$\begin{array}{r} 29 \\ \times\ 2 \\ \hline \end{array}$$

(7)
$$\begin{array}{r} 28 \\ \times\ 3 \\ \hline \end{array}$$

(2)
$$\begin{array}{r} 28 \\ \times\ 2 \\ \hline \end{array}$$

(5)
$$\begin{array}{r} 25 \\ \times\ 3 \\ \hline \end{array}$$

(8)
$$\begin{array}{r} 26 \\ \times\ 3 \\ \hline \end{array}$$

(3)
$$\begin{array}{r} 27 \\ \times\ 2 \\ \hline \end{array}$$

(6)
$$\begin{array}{r} 24 \\ \times\ 3 \\ \hline \end{array}$$

3 Calculate.

6 points per question

(1)
$$\begin{array}{r} 37 \\ \times\ 2 \\ \hline \end{array}$$

(3)
$$\begin{array}{r} 38 \\ \times\ 2 \\ \hline \end{array}$$

(5)
$$\begin{array}{r} 48 \\ \times\ 2 \\ \hline \end{array}$$

(2)
$$\begin{array}{r} 36 \\ \times\ 2 \\ \hline \end{array}$$

(4)
$$\begin{array}{r} 49 \\ \times\ 2 \\ \hline \end{array}$$

Don't forget to carry over!

2-Digits×1-Digit 3

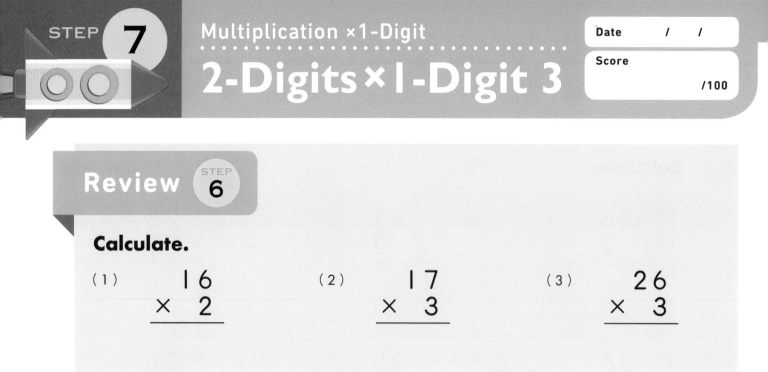

Review STEP 6

Calculate.

(1)
```
  16
×  2
```

(2)
```
  17
×  3
```

(3)
```
  26
×  3
```

1 Calculate.

5 points per question

Example ● How to calculate 18×4.

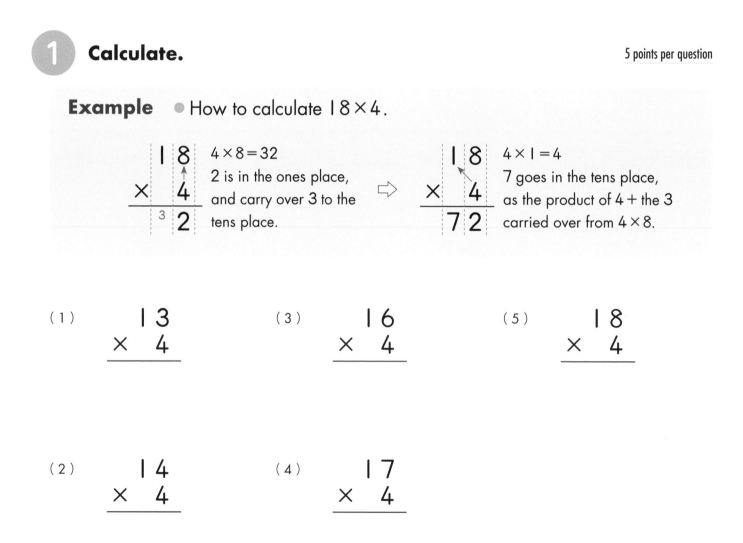

```
  18        4×8=32
×  4        2 is in the ones place,
 3 2        and carry over 3 to the
            tens place.
```
⇨
```
  18        4×1=4
×  4        7 goes in the tens place,
 72         as the product of 4 + the 3
            carried over from 4×8.
```

(1)
```
  13
×  4
```

(3)
```
  16
×  4
```

(5)
```
  18
×  4
```

(2)
```
  14
×  4
```

(4)
```
  17
×  4
```

2 Calculate.

5 points per question

(1)
$$
\begin{array}{r}
13 \\
\times\ 5 \\
\hline
\end{array}
$$

(2)
$$
\begin{array}{r}
15 \\
\times\ 5 \\
\hline
\end{array}
$$

(3)
$$
\begin{array}{r}
17 \\
\times\ 5 \\
\hline
\end{array}
$$

(4)
$$
\begin{array}{r}
12 \\
\times\ 6 \\
\hline
\end{array}
$$

(5)
$$
\begin{array}{r}
13 \\
\times\ 6 \\
\hline
\end{array}
$$

(6)
$$
\begin{array}{r}
14 \\
\times\ 6 \\
\hline
\end{array}
$$

(7)
$$
\begin{array}{r}
19 \\
\times\ 5 \\
\hline
\end{array}
$$

(8)
$$
\begin{array}{r}
19 \\
\times\ 4 \\
\hline
\end{array}
$$

(9)
$$
\begin{array}{r}
16 \\
\times\ 6 \\
\hline
\end{array}
$$

(10)
$$
\begin{array}{r}
23 \\
\times\ 4 \\
\hline
\end{array}
$$

(11)
$$
\begin{array}{r}
24 \\
\times\ 4 \\
\hline
\end{array}
$$

(12)
$$
\begin{array}{r}
12 \\
\times\ 7 \\
\hline
\end{array}
$$

(13)
$$
\begin{array}{r}
13 \\
\times\ 7 \\
\hline
\end{array}
$$

(14)
$$
\begin{array}{r}
14 \\
\times\ 7 \\
\hline
\end{array}
$$

(15)
$$
\begin{array}{r}
12 \\
\times\ 8 \\
\hline
\end{array}
$$

Review STEP 7

Calculate.

(1)
```
   1 9
 ×   4
```

(2)
```
   1 6
 ×   6
```

(3)
```
   2 3
 ×   4
```

1 Calculate.

5 points per question

Example ● How to calculate 42×3.

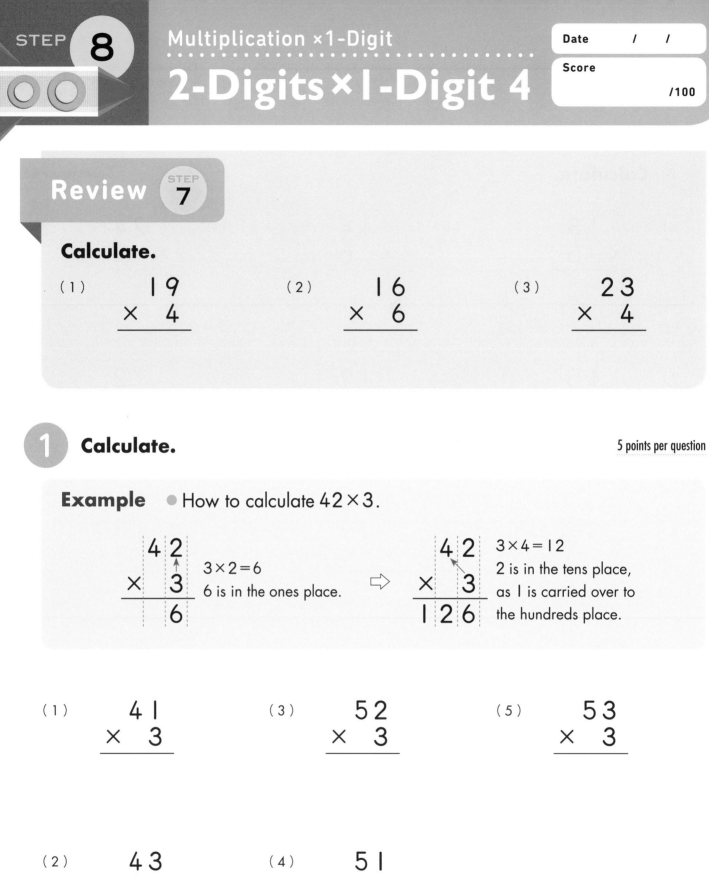

```
   4 2
 ×   3
 ───
     6
```
3×2=6
6 is in the ones place.

⇨

```
   4 2
 ×   3
 ─────
 1 2 6
```
3×4=12
2 is in the tens place,
as 1 is carried over to
the hundreds place.

(1)
```
   4 1
 ×   3
```

(3)
```
   5 2
 ×   3
```

(5)
```
   5 3
 ×   3
```

(2)
```
   4 3
 ×   3
```

(4)
```
   5 1
 ×   3
```

2 Calculate.

5 points per question

(1)
$$63 \times 2$$

(2)
$$63 \times 3$$

(3)
$$64 \times 2$$

(4)
$$62 \times 3$$

(5)
$$72 \times 2$$

(6)
$$71 \times 3$$

(7)
$$72 \times 3$$

(8)
$$61 \times 3$$

(9)
$$84 \times 2$$

(10)
$$73 \times 3$$

(11)
$$71 \times 2$$

(12)
$$74 \times 2$$

(13)
$$82 \times 3$$

(14)
$$93 \times 2$$

(15)
$$93 \times 3$$

2-Digits×1-Digit 5

Date   /   /

Score   /100

Review STEP 8

Calculate.

(1)
```
    62
×    2
```

(2)
```
    41
×    3
```

(3)
```
    52
×    2
```

1 Calculate.

5 points per question

Example ● How to calculate 58×2.

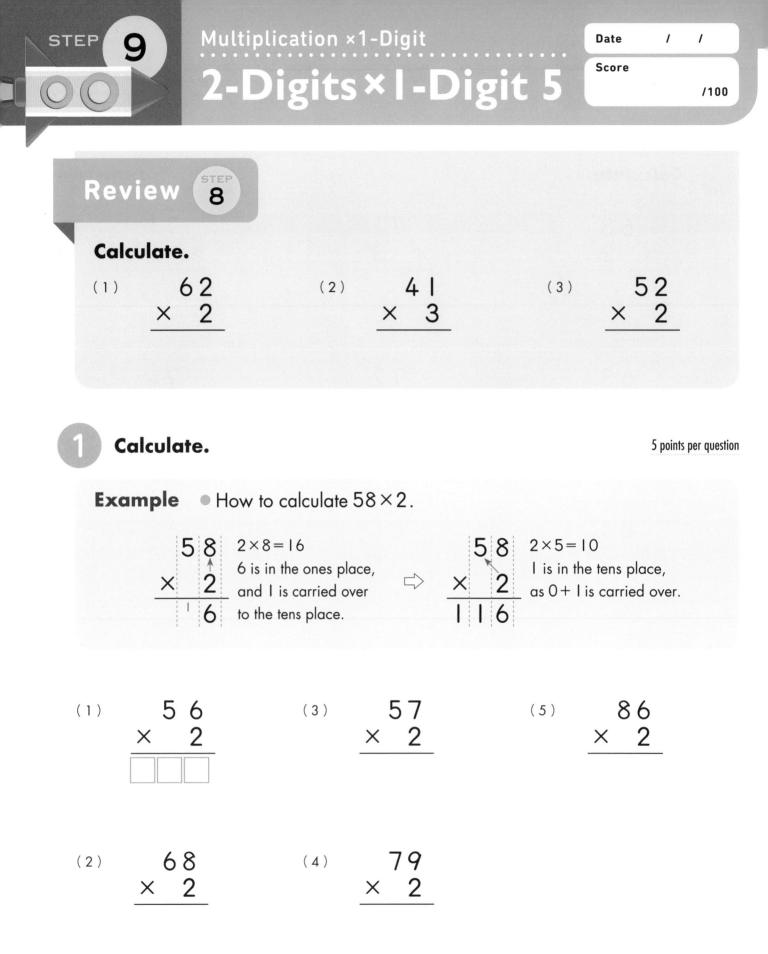

```
  58        2×8=16
×  2        6 is in the ones place,
 ¹ 6        and 1 is carried over
            to the tens place.
```
⇨
```
  58        2×5=10
×  2        1 is in the tens place,
1 1 6       as 0+1 is carried over.
```

(1)
```
    5 6
×     2
□ □ □
```

(3)
```
    5 7
×     2
```

(5)
```
    8 6
×     2
```

(2)
```
    6 8
×     2
```

(4)
```
    7 9
×     2
```

2 Calculate.

5 points per question

(1)
$$\begin{array}{r} 57 \\ \times\ 3 \\ \hline \end{array}$$

(5)
$$\begin{array}{r} 77 \\ \times\ 3 \\ \hline \end{array}$$

(9)
$$\begin{array}{r} 84 \\ \times\ 3 \\ \hline \end{array}$$

(2)
$$\begin{array}{r} 58 \\ \times\ 3 \\ \hline \end{array}$$

(6)
$$\begin{array}{r} 65 \\ \times\ 3 \\ \hline \end{array}$$

(10)
$$\begin{array}{r} 79 \\ \times\ 3 \\ \hline \end{array}$$

(3)
$$\begin{array}{r} 56 \\ \times\ 3 \\ \hline \end{array}$$

(7)
$$\begin{array}{r} 78 \\ \times\ 3 \\ \hline \end{array}$$

(11)
$$\begin{array}{r} 88 \\ \times\ 3 \\ \hline \end{array}$$

(4)
$$\begin{array}{r} 76 \\ \times\ 3 \\ \hline \end{array}$$

(8)
$$\begin{array}{r} 86 \\ \times\ 3 \\ \hline \end{array}$$

(12)
$$\begin{array}{r} 87 \\ \times\ 3 \\ \hline \end{array}$$

3 Calculate.

5 points per question

(1)
$$\begin{array}{r} 68 \\ \times\ 3 \\ \hline \end{array}$$

(2)
$$\begin{array}{r} 67 \\ \times\ 3 \\ \hline \end{array}$$

(3)
$$\begin{array}{r} 69 \\ \times\ 3 \\ \hline \end{array}$$

2-Digits×1-Digit 6

Review STEP 9

Calculate.

(1)
```
  67
×  2
```

(2)
```
  68
×  3
```

(3)
```
  56
×  2
```

1 Calculate.

5 points per question

Example ● How to calculate 34×4.

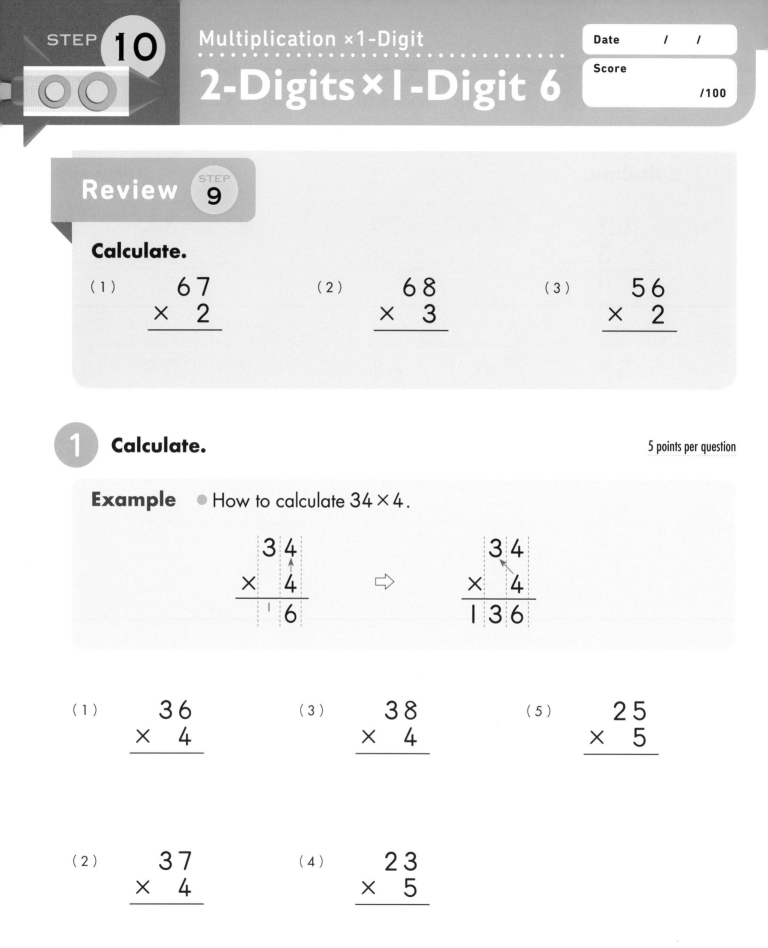

(1)
```
  36
×  4
```

(3)
```
  38
×  4
```

(5)
```
  25
×  5
```

(2)
```
  37
×  4
```

(4)
```
  23
×  5
```

2 Calculate.

5 points per question

(1)
$$\begin{array}{r} 48 \\ \times\ 5 \\ \hline \end{array}$$

(2)
$$\begin{array}{r} 56 \\ \times\ 4 \\ \hline \end{array}$$

(3)
$$\begin{array}{r} 45 \\ \times\ 4 \\ \hline \end{array}$$

(4)
$$\begin{array}{r} 26 \\ \times\ 4 \\ \hline \end{array}$$

(5)
$$\begin{array}{r} 29 \\ \times\ 4 \\ \hline \end{array}$$

(6)
$$\begin{array}{r} 63 \\ \times\ 4 \\ \hline \end{array}$$

(7)
$$\begin{array}{r} 48 \\ \times\ 4 \\ \hline \end{array}$$

(8)
$$\begin{array}{r} 23 \\ \times\ 5 \\ \hline \end{array}$$

(9)
$$\begin{array}{r} 32 \\ \times\ 5 \\ \hline \end{array}$$

(10)
$$\begin{array}{r} 79 \\ \times\ 5 \\ \hline \end{array}$$

(11)
$$\begin{array}{r} 86 \\ \times\ 4 \\ \hline \end{array}$$

(12)
$$\begin{array}{r} 86 \\ \times\ 5 \\ \hline \end{array}$$

(13)
$$\begin{array}{r} 87 \\ \times\ 5 \\ \hline \end{array}$$

(14)
$$\begin{array}{r} 76 \\ \times\ 4 \\ \hline \end{array}$$

(15)
$$\begin{array}{r} 84 \\ \times\ 5 \\ \hline \end{array}$$

2-Digits×1-Digit 7

Date   /   /

Score   /100

Review STEP 10

Calculate.

(1)
```
   37
×   4
─────
```

(2)
```
   43
×   5
─────
```

(3)
```
   27
×   4
─────
```

1 Calculate.

5 points per question

Example ● How to calculate 28×7.

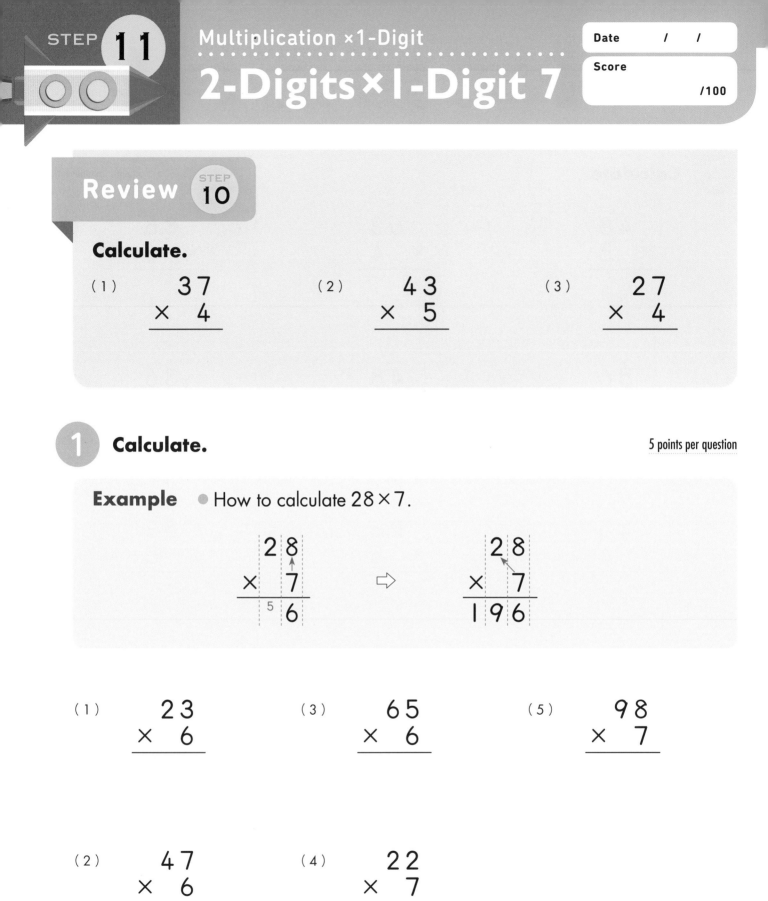

(1)
```
   23
×   6
─────
```

(3)
```
   65
×   6
─────
```

(5)
```
   98
×   7
─────
```

(2)
```
   47
×   6
─────
```

(4)
```
   22
×   7
─────
```

2 Calculate.

5 points per question

(1)
$$\begin{array}{r} 17 \\ \times\ 6 \\ \hline \end{array}$$

(2)
$$\begin{array}{r} 19 \\ \times\ 7 \\ \hline \end{array}$$

(3)
$$\begin{array}{r} 49 \\ \times\ 7 \\ \hline \end{array}$$

(4)
$$\begin{array}{r} 59 \\ \times\ 7 \\ \hline \end{array}$$

(5)
$$\begin{array}{r} 46 \\ \times\ 7 \\ \hline \end{array}$$

(6)
$$\begin{array}{r} 68 \\ \times\ 6 \\ \hline \end{array}$$

(7)
$$\begin{array}{r} 86 \\ \times\ 7 \\ \hline \end{array}$$

(8)
$$\begin{array}{r} 87 \\ \times\ 6 \\ \hline \end{array}$$

(9)
$$\begin{array}{r} 73 \\ \times\ 7 \\ \hline \end{array}$$

(10)
$$\begin{array}{r} 77 \\ \times\ 7 \\ \hline \end{array}$$

(11)
$$\begin{array}{r} 88 \\ \times\ 7 \\ \hline \end{array}$$

(12)
$$\begin{array}{r} 89 \\ \times\ 7 \\ \hline \end{array}$$

(13)
$$\begin{array}{r} 29 \\ \times\ 6 \\ \hline \end{array}$$

(14)
$$\begin{array}{r} 43 \\ \times\ 7 \\ \hline \end{array}$$

(15)
$$\begin{array}{r} 72 \\ \times\ 7 \\ \hline \end{array}$$

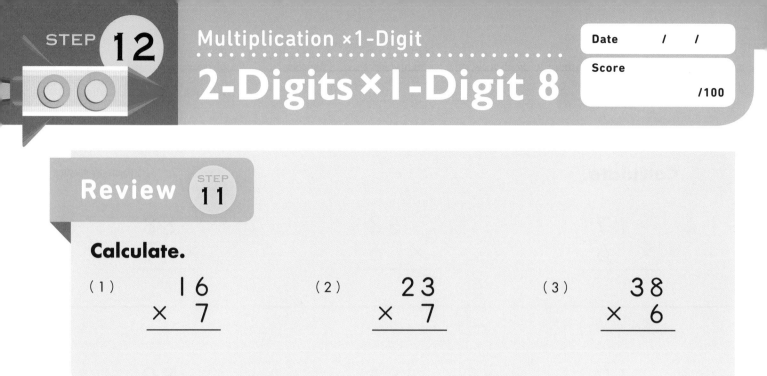

Review STEP 11

Calculate.

(1)
```
   16
×   7
```

(2)
```
   23
×   7
```

(3)
```
   38
×   6
```

1 **Calculate.**

5 points per question

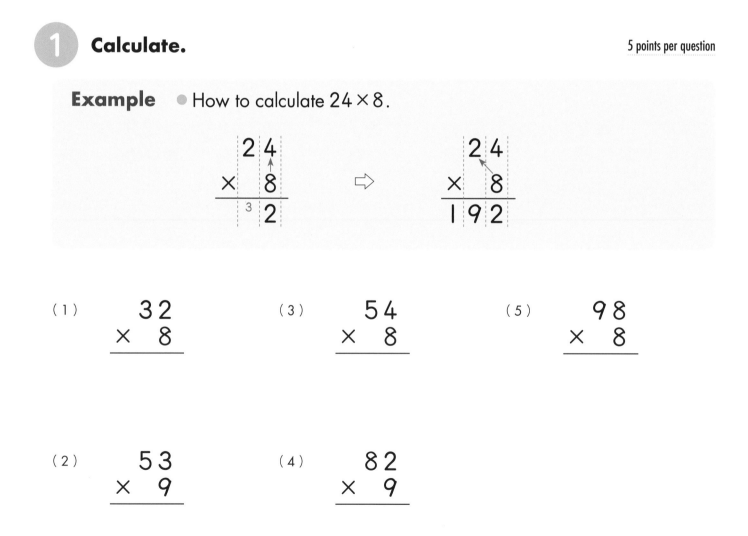

Example How to calculate 24×8.

```
   24
×   8
  ³ 2
```
⇨
```
   24
×   8
 192
```

(1)
```
   32
×   8
```

(3)
```
   54
×   8
```

(5)
```
   98
×   8
```

(2)
```
   53
×   9
```

(4)
```
   82
×   9
```

Multiplication 2× to 9×

STEP 5-16
Multiplication ×1-Digit

Multiplication ×2-Digits

Mental Math Division

Division ÷1-Digit

Division 2-Digits ÷ 2-Digits

Division 3-Digits ÷ 2-Digits

② Calculate.

5 points per question

(1)
$$35 \times 8$$

(2)
$$23 \times 9$$

(3)
$$17 \times 8$$

(4)
$$48 \times 8$$

(5)
$$37 \times 9$$

(6)
$$29 \times 9$$

(7)
$$49 \times 8$$

(8)
$$45 \times 9$$

(9)
$$65 \times 8$$

(10)
$$79 \times 9$$

(11)
$$67 \times 8$$

(12)
$$58 \times 9$$

(13)
$$38 \times 8$$

(14)
$$34 \times 9$$

(15)
$$56 \times 9$$

Multiplication ×1-Digit (2-Digits×1-Digit)

Date / /

Score /100

Review STEP 5   **Calculate.** 3 points per question

(1)
$$
\begin{array}{r}
12 \\
\times\ 4 \\
\hline
\end{array}
$$

(3)
$$
\begin{array}{r}
24 \\
\times\ 2 \\
\hline
\end{array}
$$

(5)
$$
\begin{array}{r}
22 \\
\times\ 4 \\
\hline
\end{array}
$$

(2)
$$
\begin{array}{r}
23 \\
\times\ 3 \\
\hline
\end{array}
$$

(4)
$$
\begin{array}{r}
32 \\
\times\ 3 \\
\hline
\end{array}
$$

(6)
$$
\begin{array}{r}
42 \\
\times\ 2 \\
\hline
\end{array}
$$

Review STEP 6 – STEP 8   **Calculate.** 4 points per question

(1)
$$
\begin{array}{r}
18 \\
\times\ 4 \\
\hline
\end{array}
$$

(3)
$$
\begin{array}{r}
26 \\
\times\ 2 \\
\hline
\end{array}
$$

(5)
$$
\begin{array}{r}
62 \\
\times\ 3 \\
\hline
\end{array}
$$

(2)
$$
\begin{array}{r}
15 \\
\times\ 5 \\
\hline
\end{array}
$$

(4)
$$
\begin{array}{r}
63 \\
\times\ 2 \\
\hline
\end{array}
$$

Review STEP 9   **Calculate.** 3 points per question

(1)
$$
\begin{array}{r}
56 \\
\times\ 2 \\
\hline
\end{array}
$$

(2)
$$
\begin{array}{r}
65 \\
\times\ 2 \\
\hline
\end{array}
$$

(3)
$$
\begin{array}{r}
78 \\
\times\ 3 \\
\hline
\end{array}
$$

(4)
$$
\begin{array}{r}
89 \\
\times\ 3 \\
\hline
\end{array}
$$

Review STEP **10** **Calculate.**   3 points per question

(1)
$$37 \times 5$$

(3)
$$58 \times 4$$

(5)
$$46 \times 5$$

(2)
$$34 \times 5$$

(4)
$$69 \times 4$$

(6)
$$87 \times 4$$

Review STEP **11** **Calculate.**   4 points per question

(1)
$$25 \times 7$$

(3)
$$64 \times 7$$

(5)
$$52 \times 7$$

(2)
$$38 \times 6$$

(4)
$$78 \times 6$$

Review STEP **12** **Calculate.**   3 points per question

(1)
$$52 \times 8$$

(2)
$$69 \times 9$$

(3)
$$36 \times 8$$

(4)
$$43 \times 9$$

3-Digits×1-Digit 1

Date   /   /

Score   /100

Review STEP 12

Calculate.

(1)
$$\begin{array}{r} 54 \\ \times\ \ 8 \\ \hline \end{array}$$

(2)
$$\begin{array}{r} 47 \\ \times\ \ 9 \\ \hline \end{array}$$

(3)
$$\begin{array}{r} 39 \\ \times\ \ 8 \\ \hline \end{array}$$

---

1 **Calculate.**

8 points per question

Example ● How to calculate 213×3.

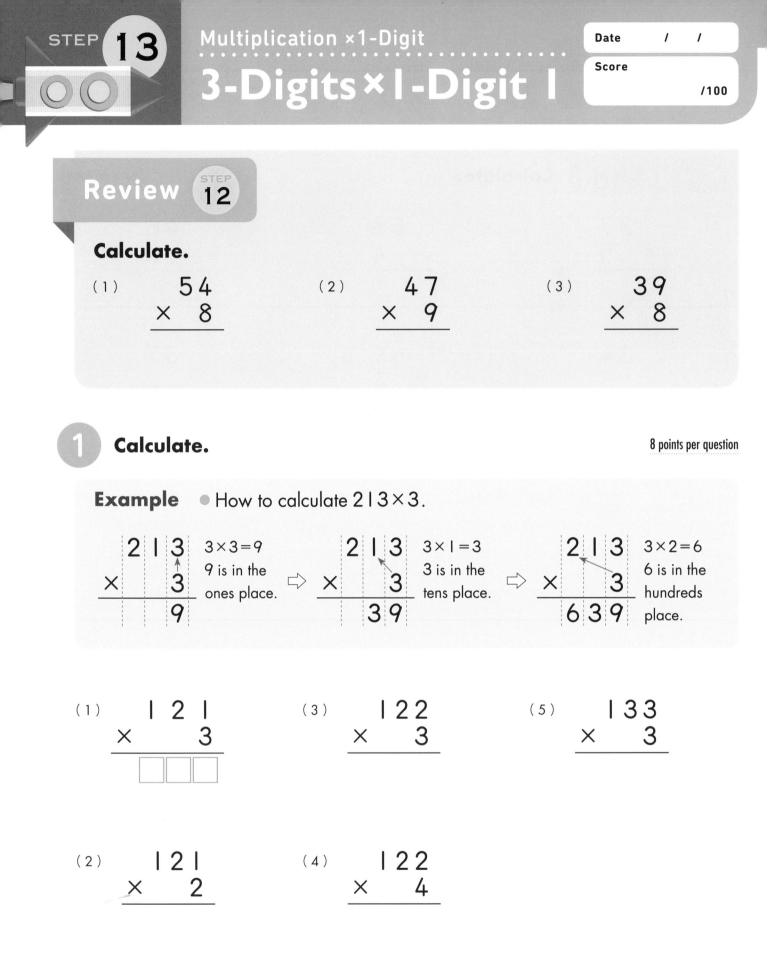

$$\begin{array}{r} 213 \\ \times\ \ \ \ 3 \\ \hline 9 \end{array}$$
3×3=9
9 is in the ones place.

⇨

$$\begin{array}{r} 213 \\ \times\ \ \ \ 3 \\ \hline 39 \end{array}$$
3×1=3
3 is in the tens place.

⇨

$$\begin{array}{r} 213 \\ \times\ \ \ \ 3 \\ \hline 639 \end{array}$$
3×2=6
6 is in the hundreds place.

(1)
$$\begin{array}{r} 121 \\ \times\ \ \ \ 3 \\ \hline \square\square\square \end{array}$$

(3)
$$\begin{array}{r} 122 \\ \times\ \ \ \ 3 \\ \hline \end{array}$$

(5)
$$\begin{array}{r} 133 \\ \times\ \ \ \ 3 \\ \hline \end{array}$$

(2)
$$\begin{array}{r} 121 \\ \times\ \ \ \ 2 \\ \hline \end{array}$$

(4)
$$\begin{array}{r} 122 \\ \times\ \ \ \ 4 \\ \hline \end{array}$$

2 Calculate.

5 points per question

(1)
$$212 \times 3$$

(2)
$$221 \times 3$$

(3)
$$222 \times 2$$

(4)
$$221 \times 4$$

(5)
$$234 \times 2$$

(6)
$$233 \times 3$$

(7)
$$314 \times 2$$

(8)
$$323 \times 3$$

(9)
$$342 \times 2$$

(10)
$$201 \times 4$$

(11)
$$203 \times 3$$

(12)
$$101 \times 9$$

They are all calculations without carrying over.

3-Digits×1-Digit 2

Review STEP 13

Calculate.

(1)
$$\begin{array}{r} 132 \\ \times 2 \\ \hline \end{array}$$

(2)
$$\begin{array}{r} 222 \\ \times 4 \\ \hline \end{array}$$

(3)
$$\begin{array}{r} 101 \\ \times 6 \\ \hline \end{array}$$

1 Calculate.

5 points per question

Example ● How to calculate 326×2.

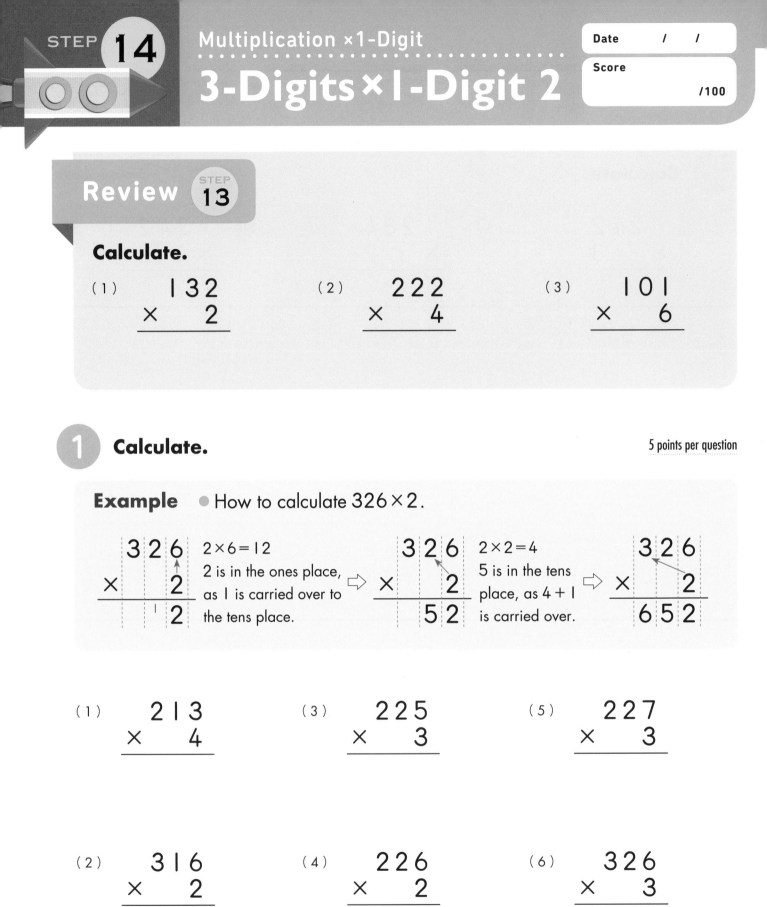

$$\begin{array}{r} 326 \\ \times 2 \\ \hline {}^{1}2 \end{array}$$
2×6=12
2 is in the ones place, as 1 is carried over to the tens place.

⇒

$$\begin{array}{r} 326 \\ \times 2 \\ \hline 52 \end{array}$$
2×2=4
5 is in the tens place, as 4+1 is carried over.

⇒

$$\begin{array}{r} 326 \\ \times 2 \\ \hline 652 \end{array}$$

(1)
$$\begin{array}{r} 213 \\ \times 4 \\ \hline \end{array}$$

(3)
$$\begin{array}{r} 225 \\ \times 3 \\ \hline \end{array}$$

(5)
$$\begin{array}{r} 227 \\ \times 3 \\ \hline \end{array}$$

(2)
$$\begin{array}{r} 316 \\ \times 2 \\ \hline \end{array}$$

(4)
$$\begin{array}{r} 226 \\ \times 2 \\ \hline \end{array}$$

(6)
$$\begin{array}{r} 326 \\ \times 3 \\ \hline \end{array}$$

2 Calculate.

5 points per question

(1)
$$\begin{array}{r} 143 \\ \times \quad 4 \\ \hline \end{array}$$

(2)
$$\begin{array}{r} 154 \\ \times \quad 3 \\ \hline \end{array}$$

(3)
$$\begin{array}{r} 167 \\ \times \quad 2 \\ \hline \end{array}$$

(4)
$$\begin{array}{r} 245 \\ \times \quad 3 \\ \hline \end{array}$$

(5)
$$\begin{array}{r} 248 \\ \times \quad 4 \\ \hline \end{array}$$

(6)
$$\begin{array}{r} 256 \\ \times \quad 2 \\ \hline \end{array}$$

(7)
$$\begin{array}{r} 367 \\ \times \quad 2 \\ \hline \end{array}$$

(8)
$$\begin{array}{r} 326 \\ \times \quad 3 \\ \hline \end{array}$$

(9)
$$\begin{array}{r} 133 \\ \times \quad 5 \\ \hline \end{array}$$

(10)
$$\begin{array}{r} 104 \\ \times \quad 5 \\ \hline \end{array}$$

(11)
$$\begin{array}{r} 143 \\ \times \quad 6 \\ \hline \end{array}$$

(12)
$$\begin{array}{r} 138 \\ \times \quad 7 \\ \hline \end{array}$$

(13)
$$\begin{array}{r} 124 \\ \times \quad 8 \\ \hline \end{array}$$

(14)
$$\begin{array}{r} 106 \\ \times \quad 9 \\ \hline \end{array}$$

Don't forget to carry over!

3-Digits×1-Digit 3

Review STEP 14

Calculate.

(1)
```
  2 1 3
×     4
```

(2)
```
  1 5 6
×     3
```

(3)
```
  1 0 4
×     6
```

1 Calculate.

5 points per question

Example ● How to calculate 812×4.

```
  8 1 2
×     4
      8
```
⇨
```
  8 1 2
×     4
    4 8
```
⇨
```
  8 1 2
×     4
3 2 4 8
```

This is a calculation whose answer will be four-digit number.

(1)
```
  3 1 2
×     4
□ □ □ □
```

(2)
```
  5 1 2
×     3
```

(3)
```
  4 1 1
×     6
```

2 Calculate.

5 points per question

(1)
```
  2 6 1
×     5
```

(2)
```
  4 9 2
×     4
```

(3)
```
  3 2 1
×     6
```

3 Calculate.

5 points per question

(1)
$$\begin{array}{r} 388 \\ \times\ \ \ 4 \\ \hline \end{array}$$

(4)
$$\begin{array}{r} 583 \\ \times\ \ \ 6 \\ \hline \end{array}$$

(7)
$$\begin{array}{r} 937 \\ \times\ \ \ 4 \\ \hline \end{array}$$

(2)
$$\begin{array}{r} 397 \\ \times\ \ \ 5 \\ \hline \end{array}$$

(5)
$$\begin{array}{r} 694 \\ \times\ \ \ 5 \\ \hline \end{array}$$

(8)
$$\begin{array}{r} 999 \\ \times\ \ \ 8 \\ \hline \end{array}$$

(3)
$$\begin{array}{r} 478 \\ \times\ \ \ 6 \\ \hline \end{array}$$

(6)
$$\begin{array}{r} 748 \\ \times\ \ \ 9 \\ \hline \end{array}$$

4 Calculate.

6 points per question

(1)
$$\begin{array}{r} 254 \\ \times\ \ \ 4 \\ \hline \end{array}$$

(3)
$$\begin{array}{r} 232 \\ \times\ \ \ 9 \\ \hline \end{array}$$

(5)
$$\begin{array}{r} 875 \\ \times\ \ \ 7 \\ \hline \end{array}$$

(2)
$$\begin{array}{r} 343 \\ \times\ \ \ 3 \\ \hline \end{array}$$

(4)
$$\begin{array}{r} 479 \\ \times\ \ \ 7 \\ \hline \end{array}$$

4-Digits×1-Digit

Review STEP 15

Calculate.

(1)
```
   4 5 3
 ×     5
```

(2)
```
   5 7 9
 ×     4
```

(3)
```
   3 4 7
 ×     6
```

1 Calculate.

5 points per question

Example ● How to calculate 2347×6.

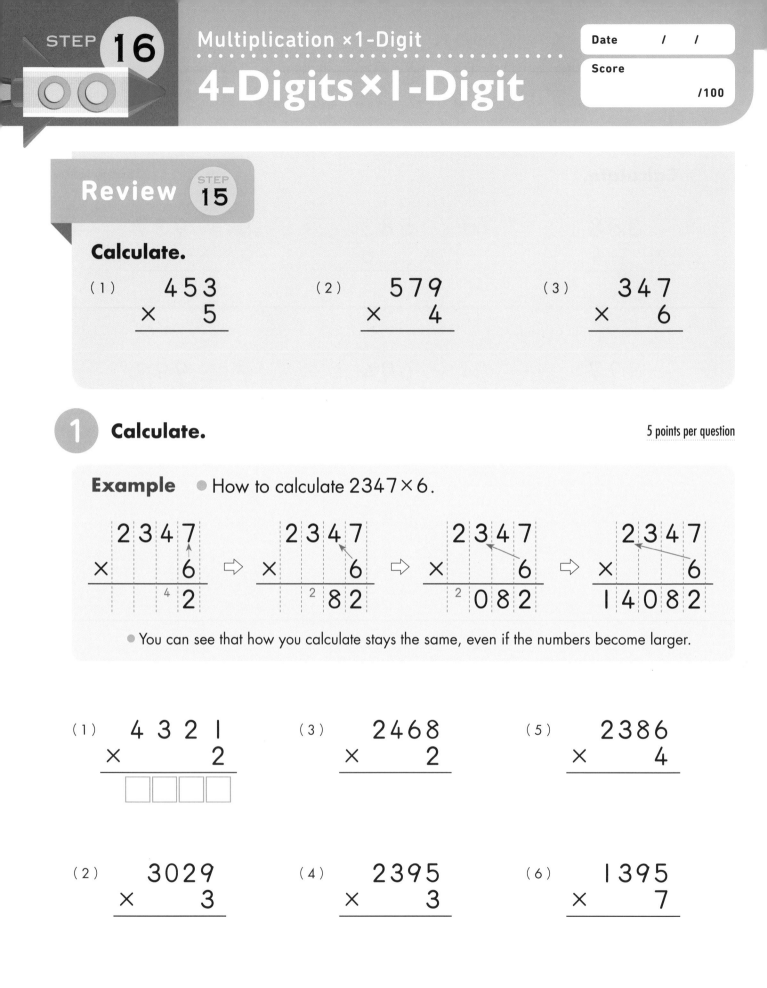

● You can see that how you calculate stays the same, even if the numbers become larger.

(1)
```
   4 3 2 1
 ×       2
```

(3)
```
   2 4 6 8
 ×       2
```

(5)
```
   2 3 8 6
 ×       4
```

(2)
```
   3 0 2 9
 ×       3
```

(4)
```
   2 3 9 5
 ×       3
```

(6)
```
   1 3 9 5
 ×       7
```

2 Calculate.

5 points per question

(1)
$$
\begin{array}{r}
4321 \\
\times \quad 4 \\
\hline \square\square\square\square\square
\end{array}
$$

(2)
$$
\begin{array}{r}
3432 \\
\times \quad 3 \\
\hline
\end{array}
$$

(3)
$$
\begin{array}{r}
5614 \\
\times \quad 4 \\
\hline
\end{array}
$$

(4)
$$
\begin{array}{r}
5143 \\
\times \quad 7 \\
\hline
\end{array}
$$

(5)
$$
\begin{array}{r}
6512 \\
\times \quad 6 \\
\hline
\end{array}
$$

(6)
$$
\begin{array}{r}
4268 \\
\times \quad 6 \\
\hline
\end{array}
$$

(7)
$$
\begin{array}{r}
2534 \\
\times \quad 5 \\
\hline
\end{array}
$$

(8)
$$
\begin{array}{r}
3785 \\
\times \quad 4 \\
\hline
\end{array}
$$

(9)
$$
\begin{array}{r}
4653 \\
\times \quad 6 \\
\hline
\end{array}
$$

(10)
$$
\begin{array}{r}
5376 \\
\times \quad 7 \\
\hline
\end{array}
$$

(11)
$$
\begin{array}{r}
5427 \\
\times \quad 8 \\
\hline
\end{array}
$$

(12)
$$
\begin{array}{r}
6845 \\
\times \quad 7 \\
\hline
\end{array}
$$

(13)
$$
\begin{array}{r}
7386 \\
\times \quad 8 \\
\hline
\end{array}
$$

(14)
$$
\begin{array}{r}
9999 \\
\times \quad 9 \\
\hline
\end{array}
$$

Be careful when carrying over.

Multiplication ×1-Digit
(3-Digits×1-Digit, 4-Digits×1-Digit)

Date / /

Score /100

Review STEP 13 **Calculate.**

4 points per question

(1)
```
  1 3 2
×     3
```

(3)
```
  2 0 3
×     2
```

(5)
```
  3 2 3
×     3
```

(2)
```
  1 4 3
×     2
```

(4)
```
  2 2 1
×     4
```

Review STEP 14 **Calculate.**

4 points per question

(1)
```
  1 1 9
×     3
```

(3)
```
  3 1 6
×     3
```

(5)
```
  1 2 4
×     5
```

(2)
```
  2 2 3
×     4
```

(4)
```
  2 4 3
×     4
```

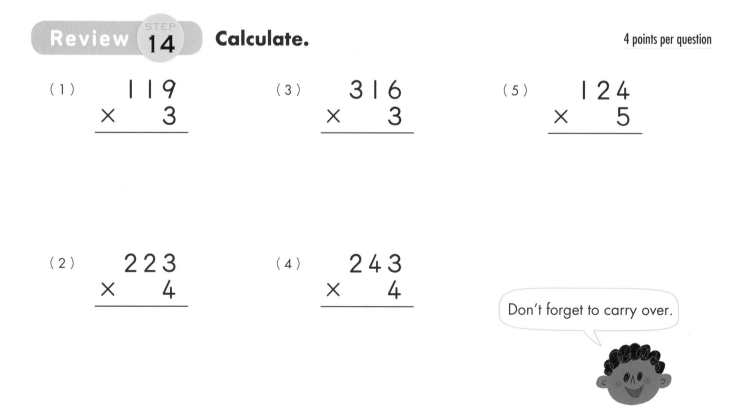

Don't forget to carry over.

Review STEP 15 STEP 16   Calculate.

4 points per question

(1)
$$315 \times 5$$

(2)
$$316 \times 4$$

(3)
$$483 \times 3$$

(4)
$$598 \times 5$$

(5)
$$635 \times 8$$

(6)
$$454 \times 6$$

(7)
$$859 \times 7$$

(8)
$$959 \times 9$$

(9)
$$924 \times 8$$

(10)
$$1285 \times 5$$

(11)
$$4439 \times 2$$

(12)
$$6812 \times 7$$

(13)
$$2691 \times 3$$

(14)
$$9927 \times 6$$

(15)
$$8319 \times 9$$

2-Digits×2-Digits 1

Date   /   /

Score   /100

Review STEP 16

Calculate.

(1)
```
  2463
×    2
```

(2)
```
  4674
×    6
```

---

1 **Calculate.**

6 points per question

Example ● How to calculate 14×20.

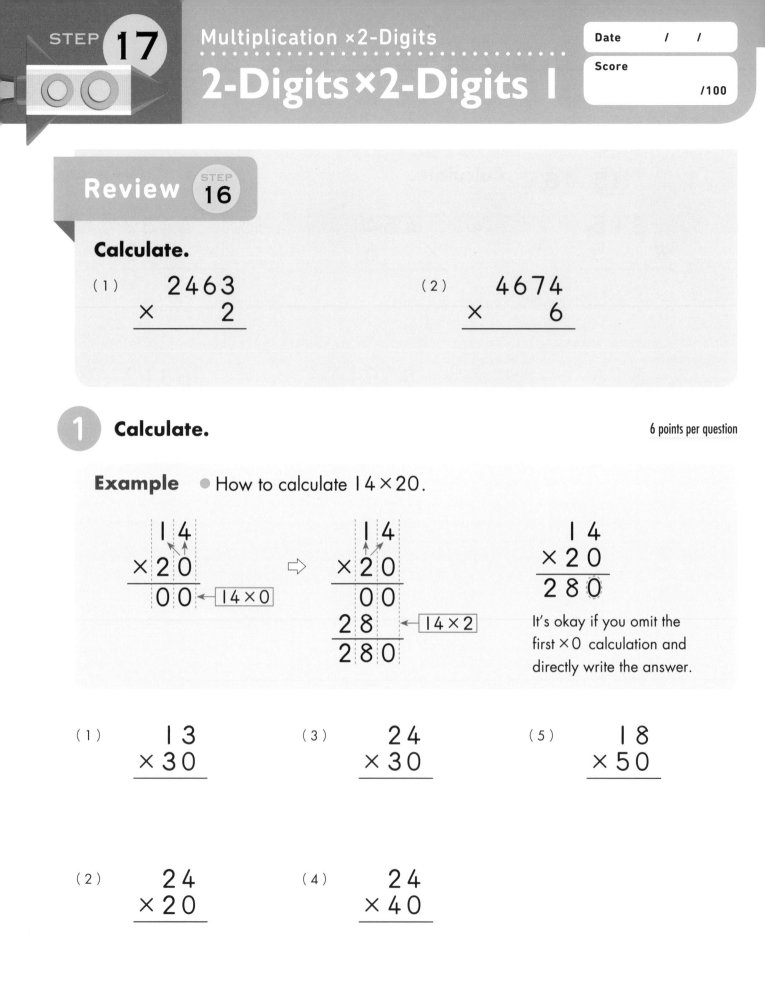

It's okay if you omit the first ×0 calculation and directly write the answer.

(1)
```
  13
×30
```

(3)
```
  24
×30
```

(5)
```
  18
×50
```

(2)
```
  24
×20
```

(4)
```
  24
×40
```

2 **Calculate.**

7 points per question

(1)
$$\begin{array}{r} 35 \\ \times\,40 \\ \hline \end{array}$$

(5)
$$\begin{array}{r} 35 \\ \times\,30 \\ \hline \end{array}$$

(8)
$$\begin{array}{r} 74 \\ \times\,80 \\ \hline \end{array}$$

(2)
$$\begin{array}{r} 57 \\ \times\,20 \\ \hline \end{array}$$

(6)
$$\begin{array}{r} 65 \\ \times\,40 \\ \hline \end{array}$$

(9)
$$\begin{array}{r} 58 \\ \times\,60 \\ \hline \end{array}$$

(3)
$$\begin{array}{r} 46 \\ \times\,40 \\ \hline \end{array}$$

(7)
$$\begin{array}{r} 35 \\ \times\,70 \\ \hline \end{array}$$

(10)
$$\begin{array}{r} 40 \\ \times\,30 \\ \hline \end{array}$$

(4)
$$\begin{array}{r} 54 \\ \times\,40 \\ \hline \end{array}$$

Don't forget to write 0 .

2-Digits×2-Digits 2

Review STEP 17

Calculate.

(1)
$$\begin{array}{r} 14 \\ \times 20 \\ \hline \end{array}$$

(2)
$$\begin{array}{r} 53 \\ \times 40 \\ \hline \end{array}$$

(3)
$$\begin{array}{r} 30 \\ \times 60 \\ \hline \end{array}$$

1 **Calculate.**

4 points per question

Example ● How to calculate 12×23.

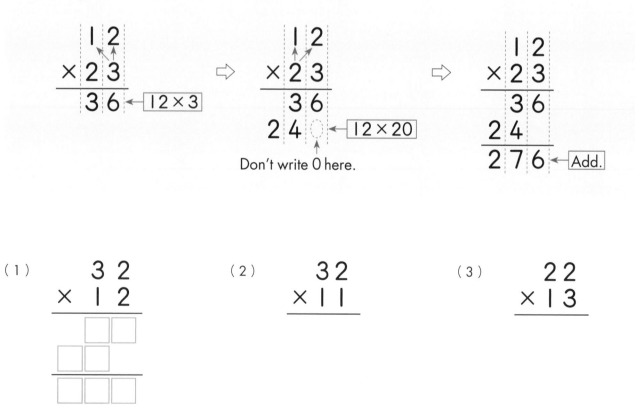

(1)
$$\begin{array}{r} 32 \\ \times 12 \\ \hline \end{array}$$

(2)
$$\begin{array}{r} 32 \\ \times 11 \\ \hline \end{array}$$

(3)
$$\begin{array}{r} 22 \\ \times 13 \\ \hline \end{array}$$

2 Calculate.

8 points per question

(1)
$$\begin{array}{r} 12 \\ \times\,24 \\ \hline \end{array}$$

(5)
$$\begin{array}{r} 42 \\ \times\,21 \\ \hline \end{array}$$

(9)
$$\begin{array}{r} 43 \\ \times\,23 \\ \hline \end{array}$$

(2)
$$\begin{array}{r} 12 \\ \times\,32 \\ \hline \end{array}$$

(6)
$$\begin{array}{r} 31 \\ \times\,27 \\ \hline \end{array}$$

(10)
$$\begin{array}{r} 53 \\ \times\,14 \\ \hline \end{array}$$

(3)
$$\begin{array}{r} 22 \\ \times\,12 \\ \hline \end{array}$$

(7)
$$\begin{array}{r} 23 \\ \times\,26 \\ \hline \end{array}$$

(11)
$$\begin{array}{r} 52 \\ \times\,18 \\ \hline \end{array}$$

(4)
$$\begin{array}{r} 41 \\ \times\,12 \\ \hline \end{array}$$

(8)
$$\begin{array}{r} 42 \\ \times\,23 \\ \hline \end{array}$$

These are calculations where you do not have to carry over when adding the numbers.

2-Digits×2-Digits 3

Date   /   /

Score   /100

Review STEP 18

Calculate.

(1)
```
    1 2
  × 2 2
```

(2)
```
    5 2
  × 1 2
```

(3)
```
    3 2
  × 1 4
```

1 Calculate.

4 points per question

Example ● How to calculate 32 × 13.

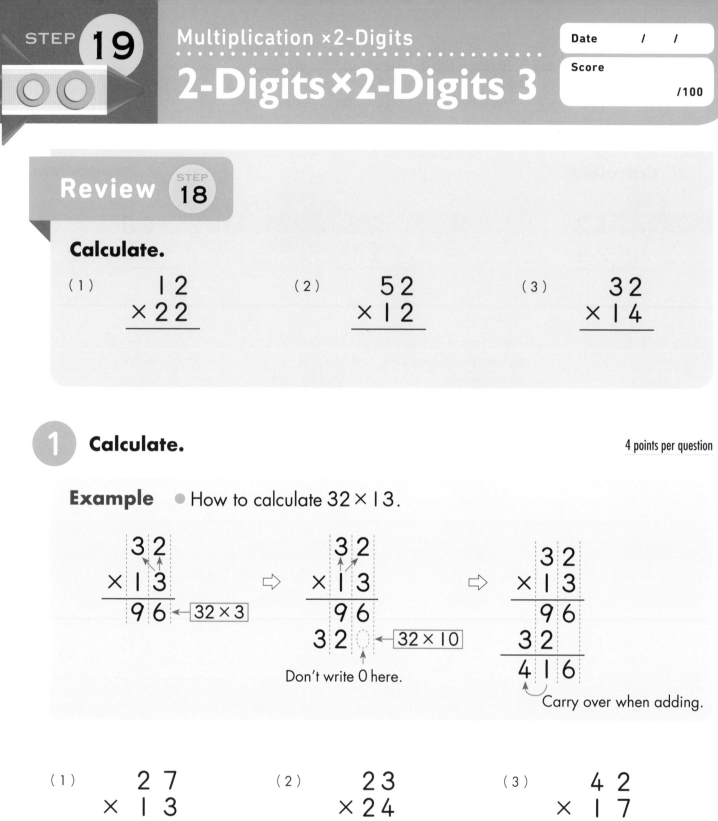

(1)
```
    2 7
  × 1 3
```

(2)
```
    2 3
  × 2 4
```

(3)
```
    4 2
  × 1 7
```

2 Calculate.

8 points per question

(1)
$$\begin{array}{r} 23 \\ \times 22 \\ \hline \end{array}$$

(5)
$$\begin{array}{r} 32 \\ \times 16 \\ \hline \end{array}$$

(9)
$$\begin{array}{r} 38 \\ \times 16 \\ \hline \end{array}$$

(2)
$$\begin{array}{r} 18 \\ \times 45 \\ \hline \end{array}$$

(6)
$$\begin{array}{r} 33 \\ \times 26 \\ \hline \end{array}$$

(10)
$$\begin{array}{r} 34 \\ \times 28 \\ \hline \end{array}$$

(3)
$$\begin{array}{r} 14 \\ \times 36 \\ \hline \end{array}$$

(7)
$$\begin{array}{r} 23 \\ \times 28 \\ \hline \end{array}$$

(11)
$$\begin{array}{r} 25 \\ \times 36 \\ \hline \end{array}$$

(4)
$$\begin{array}{r} 32 \\ \times 13 \\ \hline \end{array}$$

(8)
$$\begin{array}{r} 32 \\ \times 25 \\ \hline \end{array}$$

Don't forget to carry over!

2-Digits×2-Digits 4

Date   /   /

Score   /100

Review STEP 19

Calculate.

(1)
```
    2 8
  × 1 3
```

(2)
```
    4 5
  × 1 6
```

(3)
```
    2 3
  × 2 7
```

1 Calculate.

4 points per question

Example ● How to calculate 36×47.

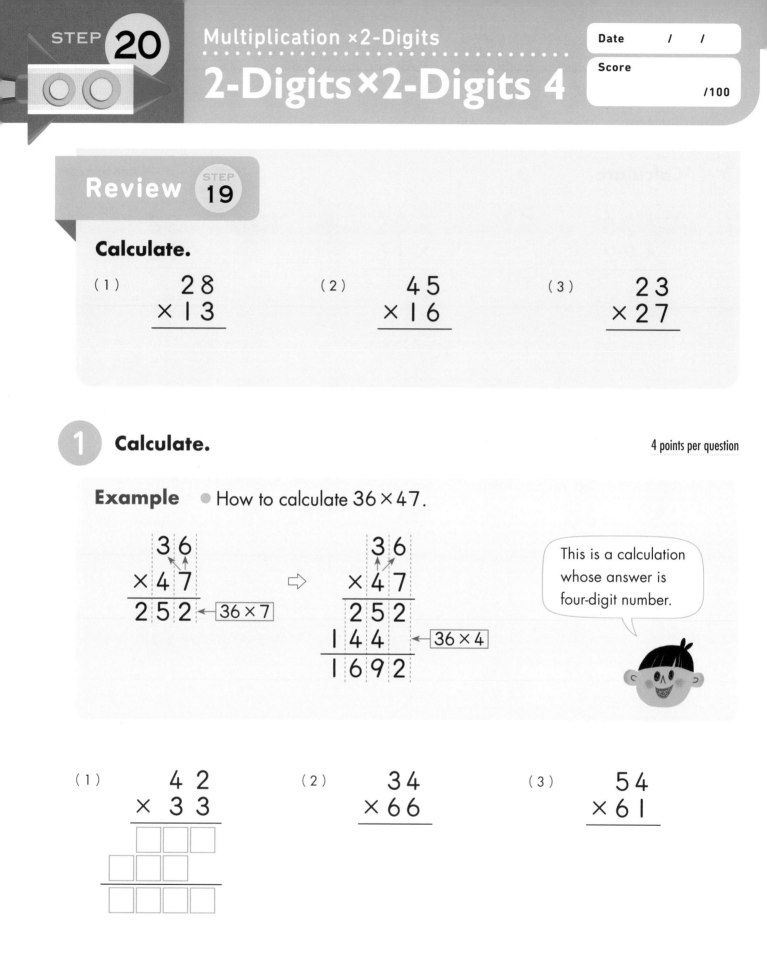

```
    3 6
  × 4 7
  2 5 2  ← 36 × 7
```
⇨
```
    3 6
  × 4 7
  2 5 2
  1 4 4  ← 36 × 4
  1 6 9 2
```

This is a calculation whose answer is four-digit number.

(1)
```
    4 2
  ×  3 3
```

(2)
```
    3 4
  × 6 6
```

(3)
```
    5 4
  × 6 1
```

2 **Calculate.**

8 points per question

(1)
$$\begin{array}{r} 54 \\ \times 82 \\ \hline \end{array}$$

(5)
$$\begin{array}{r} 58 \\ \times 92 \\ \hline \end{array}$$

(9)
$$\begin{array}{r} 79 \\ \times 86 \\ \hline \end{array}$$

(2)
$$\begin{array}{r} 46 \\ \times 99 \\ \hline \end{array}$$

(6)
$$\begin{array}{r} 64 \\ \times 84 \\ \hline \end{array}$$

(10)
$$\begin{array}{r} 88 \\ \times 87 \\ \hline \end{array}$$

(3)
$$\begin{array}{r} 82 \\ \times 56 \\ \hline \end{array}$$

(7)
$$\begin{array}{r} 75 \\ \times 87 \\ \hline \end{array}$$

(11)
$$\begin{array}{r} 89 \\ \times 98 \\ \hline \end{array}$$

(4)
$$\begin{array}{r} 65 \\ \times 82 \\ \hline \end{array}$$

(8)
$$\begin{array}{r} 78 \\ \times 83 \\ \hline \end{array}$$

They are calculations where you do not have to carry over when adding the numbers.

2-Digits×2-Digits 5

Review STEP 20

Calculate.

(1)
```
    34
  × 66
```

(2)
```
    54
  × 82
```

(3)
```
    77
  × 84
```

1 Calculate.

4 points per question

Example ● How to calculate 45×26.

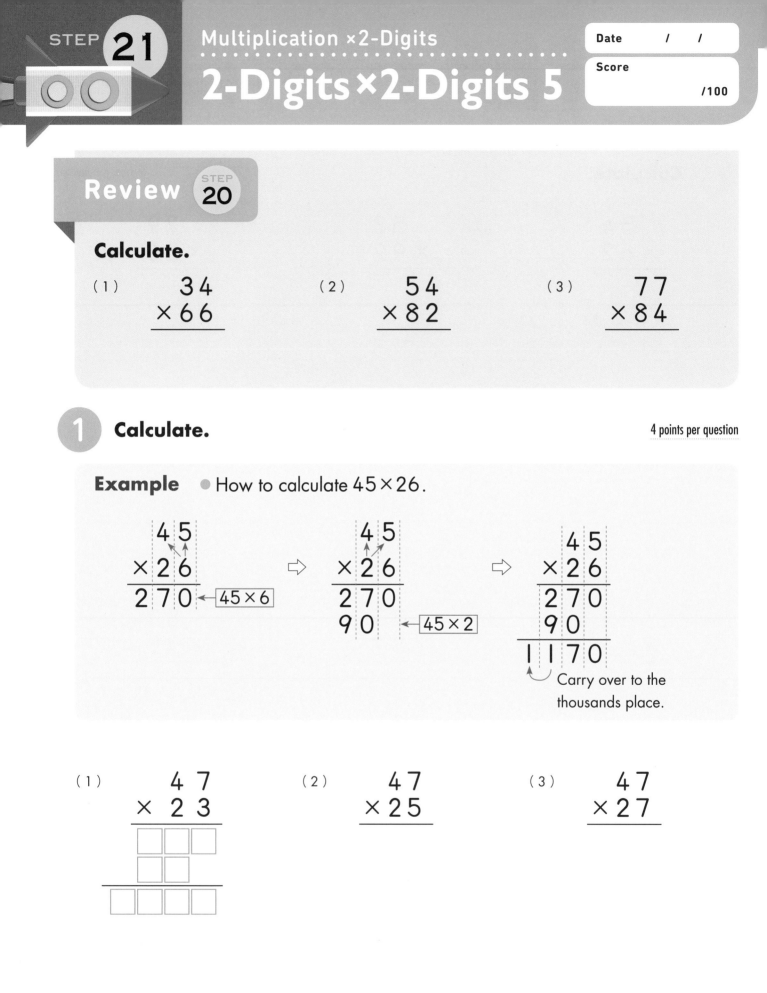

```
    4 5
  × 2 6
  2 7 0   ← 45×6
```
⇨
```
    4 5
  × 2 6
  2 7 0
    9 0   ← 45×2
```
⇨
```
    4 5
  × 2 6
  2 7 0
    9 0
  1 1 7 0
```
Carry over to the thousands place.

(1)
```
    4 7
  × 2 3
```

(2)
```
    4 7
  × 2 5
```

(3)
```
    4 7
  × 2 7
```

2 Calculate.

8 points per question

(1)
$$64 \times 18$$

(5)
$$43 \times 27$$

(9)
$$84 \times 13$$

(2)
$$76 \times 17$$

(6)
$$32 \times 37$$

(10)
$$28 \times 38$$

(3)
$$64 \times 17$$

(7)
$$84 \times 14$$

(11)
$$76 \times 14$$

(4)
$$42 \times 26$$

(8)
$$84 \times 15$$

These are calculations where the numbers are carried over to the thousands place.

2-Digits×2-Digits 6

Review STEP 21

Calculate.

(1)
```
   42
 × 26
```

(2)
```
   43
 × 27
```

(3)
```
   28
 × 39
```

1 Calculate.

4 points per question

Example ● How to calculate 32 × 69.

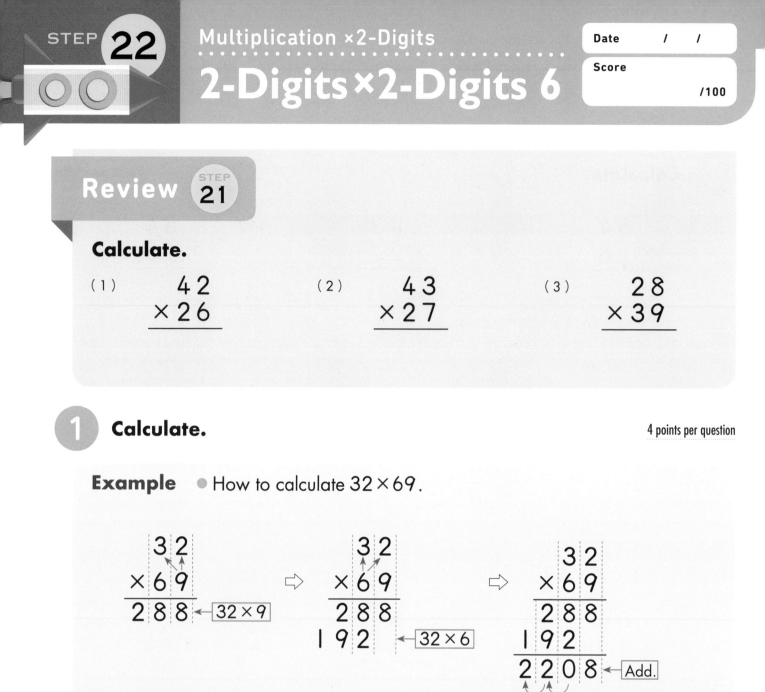

Carry over to the hundreds place and the thousands place.

(1)
```
   74
 × 28
```

(2)
```
   84
 × 27
```

(3)
```
   63
 × 39
```

2 Calculate.

8 points per question

(1)
$$\begin{array}{r} 64 \\ \times\ 49 \\ \hline \end{array}$$

(5)
$$\begin{array}{r} 86 \\ \times\ 87 \\ \hline \end{array}$$

(9)
$$\begin{array}{r} 74 \\ \times\ 89 \\ \hline \end{array}$$

(2)
$$\begin{array}{r} 62 \\ \times\ 49 \\ \hline \end{array}$$

(6)
$$\begin{array}{r} 63 \\ \times\ 89 \\ \hline \end{array}$$

(10)
$$\begin{array}{r} 86 \\ \times\ 98 \\ \hline \end{array}$$

(3)
$$\begin{array}{r} 58 \\ \times\ 69 \\ \hline \end{array}$$

(7)
$$\begin{array}{r} 57 \\ \times\ 73 \\ \hline \end{array}$$

(11)
$$\begin{array}{r} 96 \\ \times\ 97 \\ \hline \end{array}$$

(4)
$$\begin{array}{r} 65 \\ \times\ 77 \\ \hline \end{array}$$

(8)
$$\begin{array}{r} 57 \\ \times\ 74 \\ \hline \end{array}$$

Be careful as you carry over.

53

2-Digits×2-Digits 7

Date / /

Score   /100

Review STEP 22

Calculate.

(1)
```
    3 2
 ×  6 9
```

(2)
```
    6 4
 ×  4 9
```

(3)
```
    5 8
 ×  6 9
```

1 Calculate.

4 points per question

Example ● How to calculate 36×28.

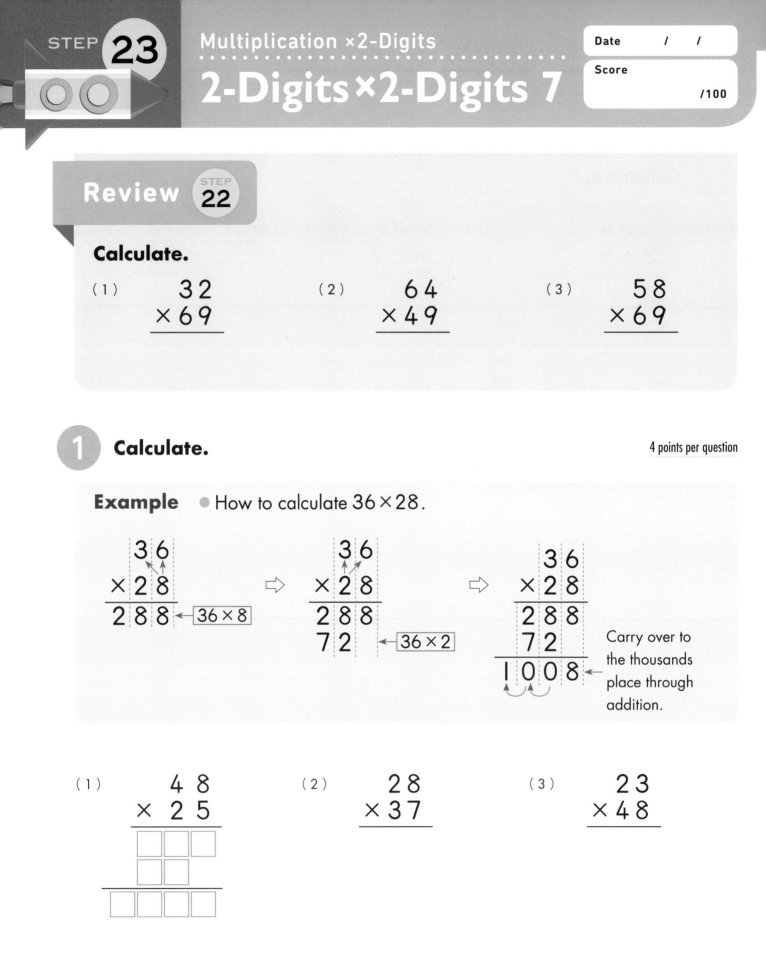

(1)
```
    4 8
 ×  2 5
```

(2)
```
    2 8
 ×  3 7
```

(3)
```
    2 3
 ×  4 8
```

2 Calculate.

8 points per question

(1)
```
   85
×  13
```

(5)
```
   48
× 26
```

(9)
```
   38
× 27
```

(2)
```
   26
× 39
```

(6)
```
   29
× 36
```

(10)
```
   53
× 19
```

(3)
```
   47
× 26
```

(7)
```
   77
× 17
```

(11)
```
   94
× 12
```

(4)
```
   86
× 12
```

(8)
```
   67
× 18
```

These are calculations whose answers will be four-digit numbers as you carry over twice.

55

3-Digits×2-Digits 1

Review STEP 23

Calculate.

(1)
```
    32
  × 35
```

(2)
```
    33
  × 38
```

(3)
```
    43
  × 29
```

1 **Calculate.**

10 points per question

Example ● How to calculate 321×12.

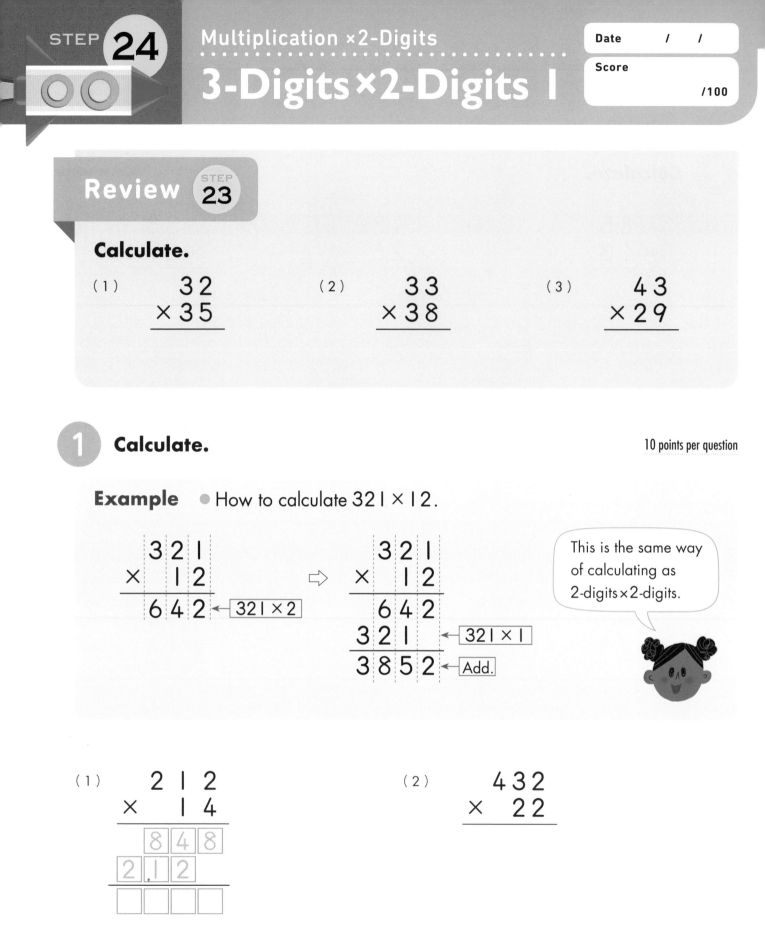

```
    321
  ×  12
    642  ← 321×2
```
⇨
```
    321
  ×  12
    642
    321    ← 321×1
   3852  ← Add.
```

This is the same way of calculating as 2-digits×2-digits.

(1)
```
    212
  ×  14
    848
   212
```

(2)
```
    432
  ×  22
```

2 Calculate.

10 points per question

(1)
$$
\begin{array}{r}
321 \\
\times \ 13 \\
\hline
\end{array}
$$

(5)
$$
\begin{array}{r}
432 \\
\times \ 14 \\
\hline
\end{array}
$$

(2)
$$
\begin{array}{r}
321 \\
\times \ 22 \\
\hline
\end{array}
$$

(6)
$$
\begin{array}{r}
346 \\
\times \ 25 \\
\hline
\end{array}
$$

(3)
$$
\begin{array}{r}
312 \\
\times \ 23 \\
\hline
\end{array}
$$

(7)
$$
\begin{array}{r}
295 \\
\times \ 32 \\
\hline
\end{array}
$$

(4)
$$
\begin{array}{r}
125 \\
\times \ 48 \\
\hline
\end{array}
$$

(8)
$$
\begin{array}{r}
402 \\
\times \ 23 \\
\hline
\end{array}
$$

3-Digits×2-Digits 2

Review STEP 24

Calculate.

(1)
```
    2 1 8
 ×   3 6
```

(2)
```
    2 9 5
 ×   3 2
```

1 Calculate.

10 points per question

Example ● How to calculate 263×48.

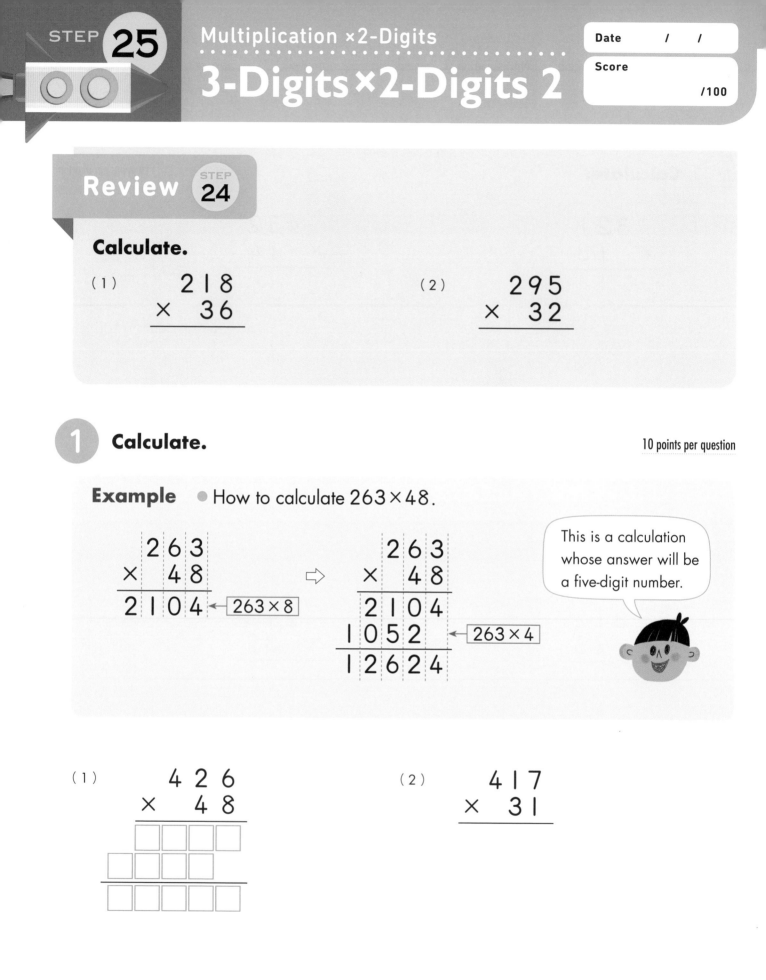

```
    2 6 3
 ×   4 8
  2 1 0 4  ← 263×8
```
⇒
```
    2 6 3
 ×   4 8
  2 1 0 4
1 0 5 2    ← 263×4
1 2 6 2 4
```

This is a calculation whose answer will be a five-digit number.

(1)
```
    4 2 6
 ×   4 8
```

(2)
```
    4 1 7
 ×   3 1
```

2 Calculate.

10 points per question

(1)
$$\begin{array}{r} 332 \\ \times\ 31 \\ \hline \end{array}$$

(5)
$$\begin{array}{r} 908 \\ \times\ 71 \\ \hline \end{array}$$

(2)
$$\begin{array}{r} 406 \\ \times\ 26 \\ \hline \end{array}$$

(6)
$$\begin{array}{r} 432 \\ \times\ 85 \\ \hline \end{array}$$

(3)
$$\begin{array}{r} 324 \\ \times\ 53 \\ \hline \end{array}$$

(7)
$$\begin{array}{r} 134 \\ \times\ 85 \\ \hline \end{array}$$

(4)
$$\begin{array}{r} 417 \\ \times\ 32 \\ \hline \end{array}$$

(8)
$$\begin{array}{r} 268 \\ \times\ 45 \\ \hline \end{array}$$

STEP **26**

Multiplication ×2-Digits

3-Digits×2-Digits 3

Date   /   /

Score

/100

Review STEP 25

Calculate.

(1)
$$
\begin{array}{r}
342 \\
\times\ \ 32 \\
\hline
\end{array}
$$

(2)
$$
\begin{array}{r}
173 \\
\times\ \ 62 \\
\hline
\end{array}
$$

1 **Calculate.**

10 points per question

Example ● How to calculate 321×32.

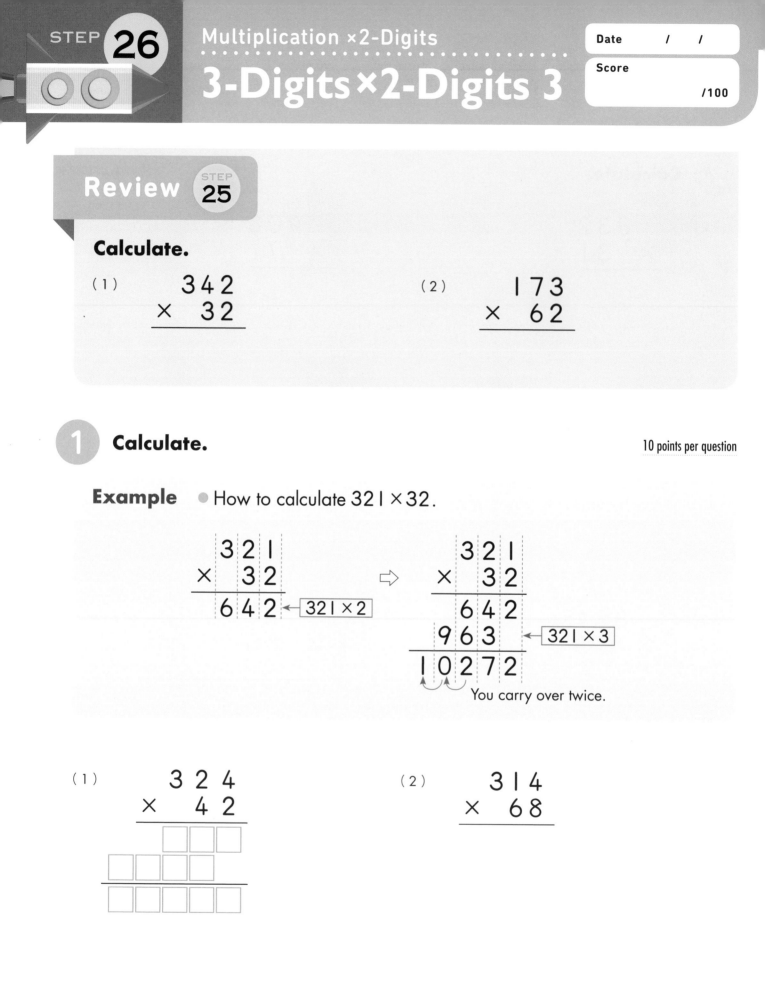

You carry over twice.

(1)
$$
\begin{array}{r}
324 \\
\times\ \ 42 \\
\hline
\end{array}
$$

(2)
$$
\begin{array}{r}
314 \\
\times\ \ 68 \\
\hline
\end{array}
$$

© Kumon Publishing Co., Ltd.

2 Calculate.

10 points per question

(1)
$$\begin{array}{r} 324 \\ \times\ 33 \\ \hline \end{array}$$

(5)
$$\begin{array}{r} 282 \\ \times\ 37 \\ \hline \end{array}$$

(2)
$$\begin{array}{r} 432 \\ \times\ 24 \\ \hline \end{array}$$

(6)
$$\begin{array}{r} 432 \\ \times\ 96 \\ \hline \end{array}$$

(3)
$$\begin{array}{r} 280 \\ \times\ 37 \\ \hline \end{array}$$

(7)
$$\begin{array}{r} 697 \\ \times\ 78 \\ \hline \end{array}$$

(4)
$$\begin{array}{r} 406 \\ \times\ 37 \\ \hline \end{array}$$

(8)
$$\begin{array}{r} 326 \\ \times\ 33 \\ \hline \end{array}$$

Multiplication ×2-Digits

Date / /

Score /100

Review STEP 17 - STEP 23 **Calculate.**

6 points per question

(1)
$$\begin{array}{r} 37 \\ \times\ 22 \\ \hline \end{array}$$

(5)
$$\begin{array}{r} 33 \\ \times\ 35 \\ \hline \end{array}$$

(8)
$$\begin{array}{r} 64 \\ \times\ 97 \\ \hline \end{array}$$

(2)
$$\begin{array}{r} 41 \\ \times\ 41 \\ \hline \end{array}$$

(6)
$$\begin{array}{r} 46 \\ \times\ 38 \\ \hline \end{array}$$

(9)
$$\begin{array}{r} 87 \\ \times\ 98 \\ \hline \end{array}$$

(3)
$$\begin{array}{r} 45 \\ \times\ 73 \\ \hline \end{array}$$

(7)
$$\begin{array}{r} 37 \\ \times\ 45 \\ \hline \end{array}$$

(10)
$$\begin{array}{r} 97 \\ \times\ 99 \\ \hline \end{array}$$

(4)
$$\begin{array}{r} 42 \\ \times\ 27 \\ \hline \end{array}$$

Be careful when carrying over.

Multiplication 2× to 9×

Multiplication ×1-Digit

STEP 17-26
Multiplication ×2-Digits

Mental Math Division

Division ÷1-Digit

Division 2-Digits ÷ 2-Digits

Division 3-Digits ÷ 2-Digits

Review STEP 24 – STEP 26   **Calculate.**

5 points per question

(1)
$$
\begin{array}{r}
321 \\
\times 21 \\
\hline
\end{array}
$$

(5)
$$
\begin{array}{r}
432 \\
\times 98 \\
\hline
\end{array}
$$

(2)
$$
\begin{array}{r}
534 \\
\times 58 \\
\hline
\end{array}
$$

(6)
$$
\begin{array}{r}
406 \\
\times 47 \\
\hline
\end{array}
$$

(3)
$$
\begin{array}{r}
806 \\
\times 61 \\
\hline
\end{array}
$$

(7)
$$
\begin{array}{r}
887 \\
\times 98 \\
\hline
\end{array}
$$

(4)
$$
\begin{array}{r}
406 \\
\times 28 \\
\hline
\end{array}
$$

(8)
$$
\begin{array}{r}
998 \\
\times 97 \\
\hline
\end{array}
$$

Review STEP 1

Calculate.

(1) $2 \times 8 =$ ☐   (3) $2 \times 6 =$ ☐   (5) $3 \times 4 =$ ☐

(2) $2 \times 7 =$ ☐   (4) $2 \times 5 =$ ☐   (6) $3 \times 5 =$ ☐

1 **Calculate.**   4 points per question

Example ● The answer for $6 \div 2$ is the same number filled in the box for $2 \times$ ☐ $= 6$.

That's inverse multiplication.

Dividend   Divisor

$2 \times \boxed{3} = 6$ $6 \div 2 = \boxed{3}$

(1) $2 \times$ ☐ $= 4$ ⟶ $4 \div 2 =$ ☐

(2) $2 \times$ ☐ $= 6$ ⟶ $6 \div 2 =$ ☐

(3) $3 \times$ ☐ $= 6$ ⟶ $6 \div 3 =$ ☐

(4) $3 \times$ ☐ $= 9$ ⟶ $9 \div 3 =$ ☐

2 **Calculate.**

4 points per question

(1) $8 \div 2 =$ ☐

(2) $10 \div 2 =$ ☐

(3) $12 \div 2 =$ ☐

(4) $18 \div 2 =$ ☐

(5) $16 \div 2 =$ ☐

(6) $14 \div 2 =$ ☐

(7) $12 \div 3 =$ ☐

(8) $15 \div 3 =$ ☐

(9) $18 \div 3 =$ ☐

(10) $27 \div 3 =$ ☐

(11) $24 \div 3 =$ ☐

(12) $21 \div 3 =$ ☐

3 **Calculate.**

4 points per question

(1) $18 \div 2 =$ ☐

(2) $24 \div 3 =$ ☐

(3) $14 \div 2 =$ ☐

(4) $21 \div 3 =$ ☐

(5) $16 \div 2 =$ ☐

(6) $27 \div 3 =$ ☐

(7) $2 \div 2 =$ ☐

(8) $3 \div 1 =$ ☐

(9) $0 \div 3 =$ 0

The answer is always 0, when 0 is divided by any number that is not 0.

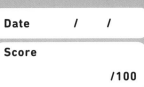

Review STEP 2 STEP 3

Calculate.

(1) $4 \times 5 =$ ☐ (3) $6 \times 7 =$ ☐ (5) $5 \times 8 =$ ☐

(2) $4 \times 6 =$ ☐ (4) $5 \times 7 =$ ☐ (6) $6 \times 9 =$ ☐

1 Calculate.

3 points per question

Example

$$4 \times \boxed{3} = 12 \qquad 12 \div 4 = \boxed{3}$$

(1) $4 \times$ ☐ $= 8 \longrightarrow 8 \div 4 =$ ☐

(2) $5 \times$ ☐ $= 15 \longrightarrow 15 \div 5 =$ ☐

(3) $6 \times$ ☐ $= 12 \longrightarrow 12 \div 6 =$ ☐

(4) $6 \times$ ☐ $= 18 \longrightarrow 18 \div 6 =$ ☐

2 Calculate.

4 points per question

(1) 16 ÷ 4 =

(2) 20 ÷ 4 =

(3) 24 ÷ 4 =

(4) 36 ÷ 4 =

(5) 32 ÷ 4 =

(6) 42 ÷ 6 =

(7) 20 ÷ 5 =

(8) 25 ÷ 5 =

(9) 30 ÷ 5 =

(10) 45 ÷ 5 =

(11) 40 ÷ 5 =

(12) 35 ÷ 5 =

(13) 24 ÷ 6 =

(14) 30 ÷ 6 =

(15) 36 ÷ 6 =

(16) 54 ÷ 6 =

(17) 48 ÷ 6 =

(18) 42 ÷ 6 =

3 Calculate.

4 points per question

(1) 6 ÷ 6 =

(2) 5 ÷ 5 =

(3) 4 ÷ 1 =

(4) 0 ÷ 6 =

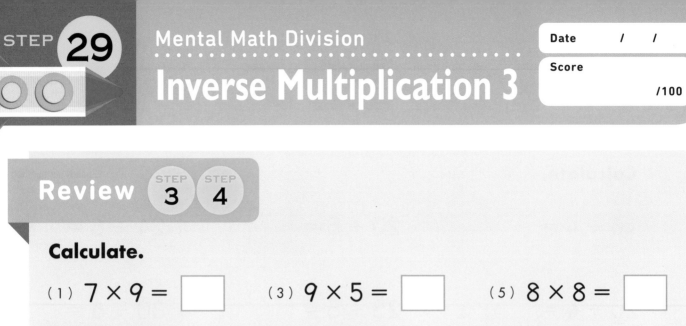

Review STEP 3 STEP 4

Calculate.

(1) $7 \times 9 = \boxed{}$

(3) $9 \times 5 = \boxed{}$

(5) $8 \times 8 = \boxed{}$

(2) $7 \times 8 = \boxed{}$

(4) $8 \times 6 = \boxed{}$

(6) $9 \times 9 = \boxed{}$

1 **Calculate.**

3 points per question

Example

$$7 \times \boxed{2} = 14 \qquad 14 \div 7 = \boxed{2}$$

(1) $7 \times \boxed{} = 21 \longrightarrow 21 \div 7 = \boxed{}$

(2) $8 \times \boxed{} = 24 \longrightarrow 24 \div 8 = \boxed{}$

(3) $9 \times \boxed{} = 18 \longrightarrow 18 \div 9 = \boxed{}$

(4) $9 \times \boxed{} = 27 \longrightarrow 27 \div 9 = \boxed{}$

2 **Calculate.**

4 points per question

(1) $28 \div 7 =$

(2) $35 \div 7 =$

(3) $42 \div 7 =$

(4) $63 \div 7 =$

(5) $56 \div 7 =$

(6) $49 \div 7 =$

(7) $32 \div 8 =$

(8) $40 \div 8 =$

(9) $48 \div 8 =$

(10) $72 \div 8 =$

(11) $64 \div 8 =$

(12) $56 \div 8 =$

(13) $36 \div 9 =$

(14) $45 \div 9 =$

(15) $54 \div 9 =$

(16) $81 \div 9 =$

(17) $72 \div 9 =$

(18) $63 \div 9 =$

3 **Calculate.**

4 points per question

(1) $7 \div 7 =$

(2) $8 \div 1 =$

(3) $9 \div 9 =$

(4) $0 \div 7 =$

STEP 30

Mental Math Division
Division with Remainders I

Date   /   /

Score
        /100

Review STEP 27

Calculate.

(1) $10 \div 2 =$ ☐   (3) $14 \div 2 =$ ☐   (5) $15 \div 3 =$ ☐

(2) $16 \div 2 =$ ☐   (4) $12 \div 2 =$ ☐   (6) $27 \div 3 =$ ☐

1 Calculate.   2 points per question

Example ● How to calculate $14 \div 3$.

$14 \div 3 =$ ☐4☐ R ☐2☐ ← Write the remainder here.

"R" means the "remainder."

A remainder is smaller than divisor.

You can find the answer by using multiplication.

(1) $4 \div 2 =$ ☐

(2) $5 \div 2 =$ ☐ R ☐

(3) $7 \div 2 =$

(4) $6 \div 3 =$ ☐

(5) $7 \div 3 =$ ☐ R ☐

(6) $10 \div 3 =$

2 Calculate.

5 points per question

(1) $8 \div 2 =$

(5) $19 \div 2 =$

(2) $9 \div 2 =$

(6) $17 \div 2 =$

(3) $13 \div 2 =$

(7) $11 \div 2 =$

(4) $15 \div 2 =$

(8) $18 \div 2 =$

3 Calculate.

6 points per question

(1) $13 \div 3 =$

(5) $19 \div 3 =$

(2) $14 \div 3 =$

(6) $20 \div 3 =$

(3) $17 \div 3 =$

(7) $28 \div 3 =$

(4) $16 \div 3 =$

(8) $26 \div 3 =$

STEP 31

Mental Math Division

Division with Remainders 2

Date / /

Score /100

Review STEP 28

Calculate.

(1) $8 \div 4 =$ ☐

(2) $24 \div 4 =$ ☐

(3) $36 \div 6 =$ ☐

(4) $10 \div 5 =$ ☐

(5) $25 \div 5 =$ ☐

(6) $45 \div 5 =$ ☐

1 Calculate.

2 points per question

Example

$9 \div 4 =$ 2 R 1

(1) $8 \div 4 =$

(2) $9 \div 4 =$

(3) $10 \div 4 =$

(4) $11 \div 4 =$

(5) $12 \div 5 =$

(6) $13 \div 5 =$

(7) $14 \div 5 =$

(8) $15 \div 5 =$

2 Calculate.

5 points per question

(1) $18 \div 4 =$

(2) $17 \div 4 =$

(3) $22 \div 4 =$

(4) $23 \div 4 =$

(5) $37 \div 4 =$

(6) $30 \div 4 =$

3 Calculate.

6 points per question

(1) $33 \div 5 =$

(2) $27 \div 5 =$

(3) $46 \div 5 =$

(4) $44 \div 5 =$

4 Calculate.

5 points per question

(1) $26 \div 6 =$

(2) $32 \div 6 =$

(3) $45 \div 6 =$

(4) $50 \div 6 =$

(5) $52 \div 6 =$

(6) $57 \div 6 =$

Review STEP 29

Calculate.

(1) $35 \div 7 =$ ☐   (3) $56 \div 8 =$ ☐   (5) $45 \div 9 =$ ☐

(2) $42 \div 7 =$ ☐   (4) $32 \div 8 =$ ☐   (6) $63 \div 9 =$ ☐

---

1   Calculate.   2 points per question

Example

$$16 \div 7 = \boxed{2} \ R \ \boxed{2}$$

(1) $20 \div 7 =$ (5) $26 \div 8 =$

(2) $21 \div 7 =$ (6) $25 \div 8 =$

(3) $22 \div 7 =$ (7) $24 \div 8 =$

(4) $23 \div 7 =$ (8) $23 \div 8 =$

2 Calculate.

6 points per question

(1) $31 \div 7 =$

(3) $52 \div 7 =$

(2) $41 \div 7 =$

(4) $65 \div 7 =$

3 Calculate.

5 points per question

(1) $36 \div 8 =$

(4) $59 \div 8 =$

(2) $43 \div 8 =$

(5) $65 \div 8 =$

(3) $57 \div 8 =$

(6) $75 \div 8 =$

4 Calculate.

5 points per question

(1) $38 \div 9 =$

(4) $57 \div 9 =$

(2) $37 \div 9 =$

(5) $75 \div 9 =$

(3) $48 \div 9 =$

(6) $83 \div 9 =$

Mental Math Division

Date   /   /

Score   /100

Review STEP 30   **Calculate.**

5 points per question

(1) $15 \div 2 =$

(2) $19 \div 3 =$

(3) $7 \div 2 =$

(4) $12 \div 3 =$

(5) $17 \div 2 =$

(6) $25 \div 3 =$

(7) $9 \div 2 =$

(8) $23 \div 3 =$

Review STEP 31   **Calculate.**

3 points per question

(1) $17 \div 4 =$

(2) $23 \div 5 =$

(3) $30 \div 4 =$

(4) $35 \div 5 =$

(5) $57 \div 6 =$

(6) $45 \div 6 =$

Review STEP **32**   **Calculate.**

3 points per question

(1) $31 \div 7 =$

(2) $36 \div 8 =$

(3) $38 \div 9 =$

(4) $53 \div 7 =$

(5) $48 \div 8 =$

(6) $58 \div 9 =$

(7) $63 \div 7 =$

(8) $65 \div 8 =$

(9) $72 \div 9 =$

(10) $67 \div 7 =$

(11) $75 \div 8 =$

(12) $83 \div 9 =$

(13) $67 \div 7 =$

(14) $77 \div 8 =$

2-Digits÷1-Digit 1

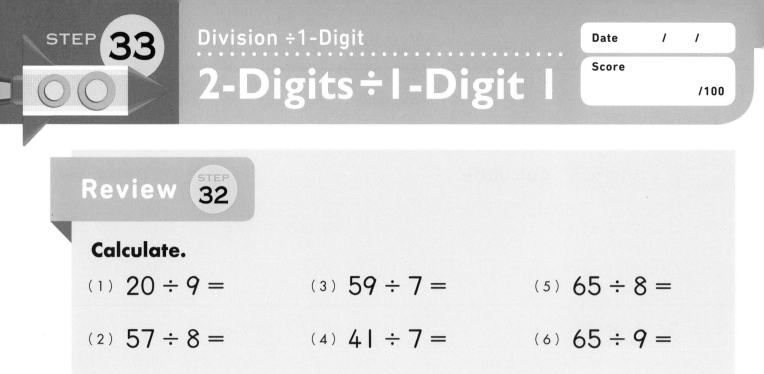

Review STEP 32

Calculate.

(1) $20 \div 9 =$

(3) $59 \div 7 =$

(5) $65 \div 8 =$

(2) $57 \div 8 =$

(4) $41 \div 7 =$

(6) $65 \div 9 =$

---

1 **Calculate.**

5 points per question

Example

● How to calculate $12 \div 2$.

```
        6  ← Write the quotient here.
  2) 1 2
     1 2  ← 2×6
        0  ← Subtract. There is no
            remainder, so write 0.
```

● How to calculate $13 \div 2$.

```
        6  ← Quotient
  2) 1 3
     1 2  ← 2×6
        1  ← Write the remainder here.
            (13−12=1)
```

Quotient means the solution of a division problem.

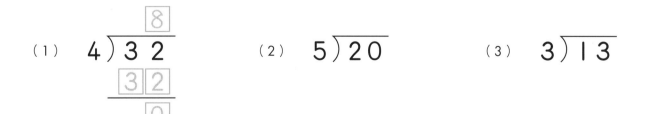

(1)
```
        8
  4) 3 2
     3 2
        0
```

(2)
```
  5) 2 0
```

(3)
```
  3) 1 3
```

© Kumon Publishing Co., Ltd.

2 Calculate.

5 points per question

(1) $2\overline{)14}$

(4) $6\overline{)24}$

(7) $7\overline{)63}$

(2) $3\overline{)21}$

(5) $5\overline{)30}$

(8) $9\overline{)45}$

(3) $4\overline{)24}$

(6) $8\overline{)32}$

3 Calculate.

9 points per question

(1) $3\overline{)20}$

(3) $5\overline{)21}$

(5) $7\overline{)62}$

(2) $6\overline{)45}$

(4) $4\overline{)35}$

2-Digits÷1-Digit 2

Review STEP 33

Calculate.

(1) 2)17

(2) 6)32

(3) 8)36

---

1 **Calculate.**

10 points per question

Example ● How to calculate $27 \div 2$.

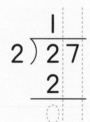

Write 1 (the answer for
$2 \div 2$) in the tens place
and subtract $2-2$.

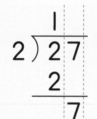

Move the 7 in the
ones place down and
calculate $7 \div 2$.

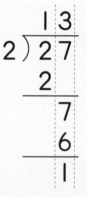

Write 3 (the answer for
$7 \div 2$) in the ones place
and subtract $7-6$.

(1) 2)24

(2) 3)37

(3) 2)31

2 Calculate.

5 points per question

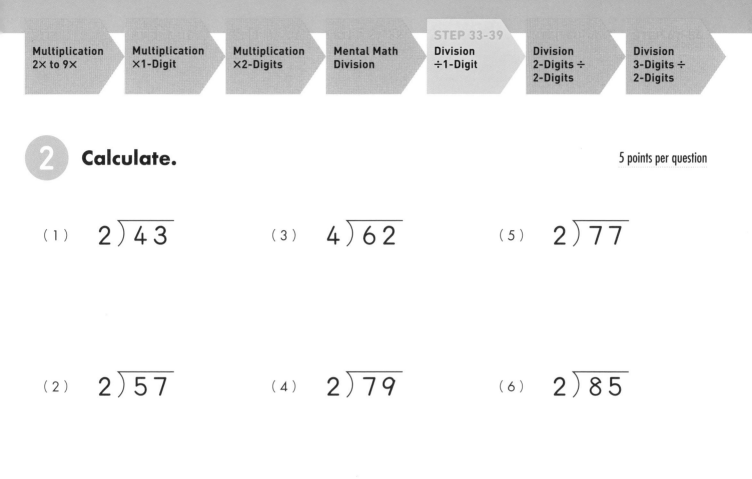

(1) 2)43

(2) 2)57

(3) 4)62

(4) 2)79

(5) 2)77

(6) 2)85

3 Calculate.

8 points per question

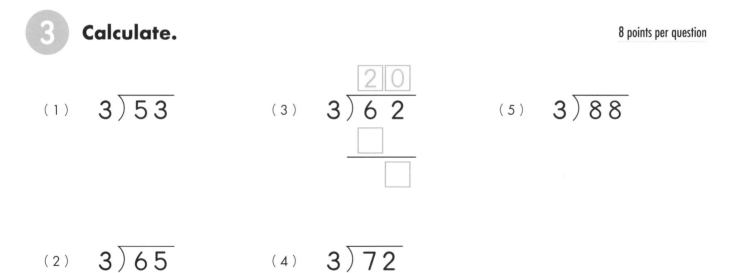

(1) 3)53

(2) 3)65

(3) 3)62   (with 2 0 above, and blank boxes below)

(4) 3)72

(5) 3)88

2-Digits÷1-Digit 3

Division ÷1-Digit

Review STEP 34

Calculate.

(1) $2\overline{)53}$

(2) $3\overline{)57}$

(3) $3\overline{)89}$

1 Calculate.

5 points per question

Example ● How to calculate $46 \div 4$.

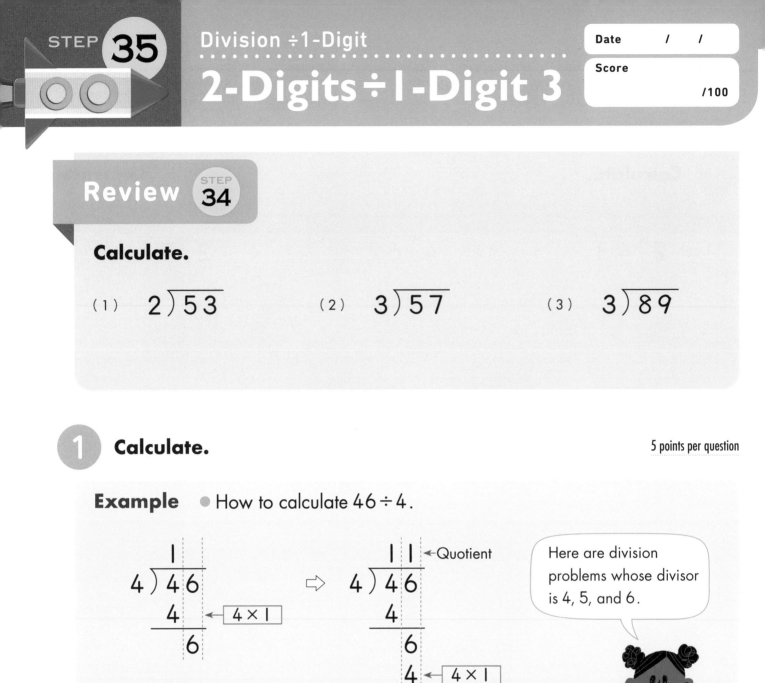

Here are division problems whose divisor is 4, 5, and 6.

(1) $4\overline{)48}$

(2) $5\overline{)56}$

2 Calculate.

10 points per question

(1) 4)65

(2) 4)69

(3) 4)73

3 Calculate.

10 points per question

(1) 5)67

(2) 5)75

(3) 5)78

4 Calculate.

10 points per question

(1) 6)74

(2) 6)81

(3) 6)99

Review STEP 35

Calculate.

(1) $4\overline{)67}$ (2) $5\overline{)79}$ (3) $6\overline{)84}$

---

1 **Calculate.**

5 points per question

Example ● How to calculate $78 \div 7$.

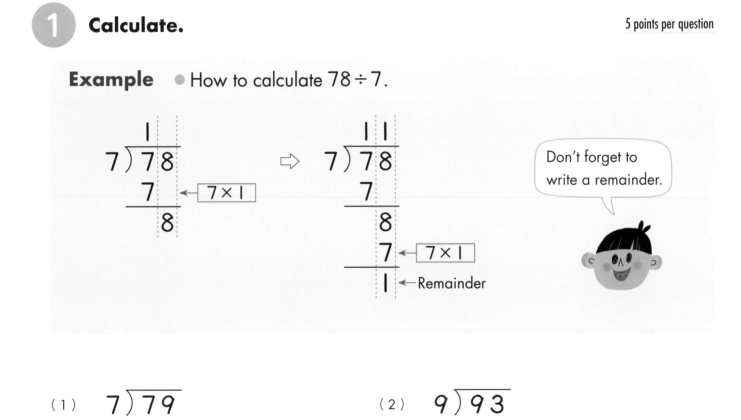

Don't forget to write a remainder.

(1) $7\overline{)79}$ (2) $9\overline{)93}$

Multiplication
2× to 9×

Multiplication
×1-Digit

Multiplication
×2-Digits

Mental Math
Division

STEP 33-39
Division
÷1-Digit

Division
2-Digits ÷
2-Digits

Division
3-Digits ÷
2-Digits

2 Calculate.
10 points per question

(1) 7)76   (2) 7)86   (3) 7)84

3 Calculate.
10 points per question

(1) 8)81   (2) 8)80   (3) 8)87

4 Calculate.
10 points per question

(1) 9)90   (2) 9)95   (3) 9)99

Review STEP 36

Calculate.

(1) $7 \overline{)85}$

(2) $8 \overline{)96}$

(3) $9 \overline{)98}$

1 Calculate.

10 points per question

Example ● How to calculate $431 \div 2$.

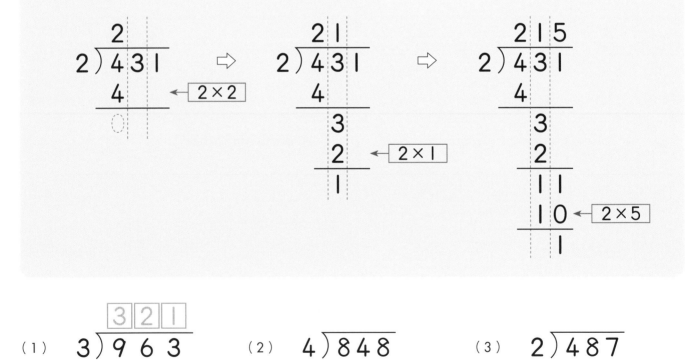

(1) $3 \overline{)963}$ — 3 2 1

(2) $4 \overline{)848}$

(3) $2 \overline{)487}$

2 Calculate.

7 points per question

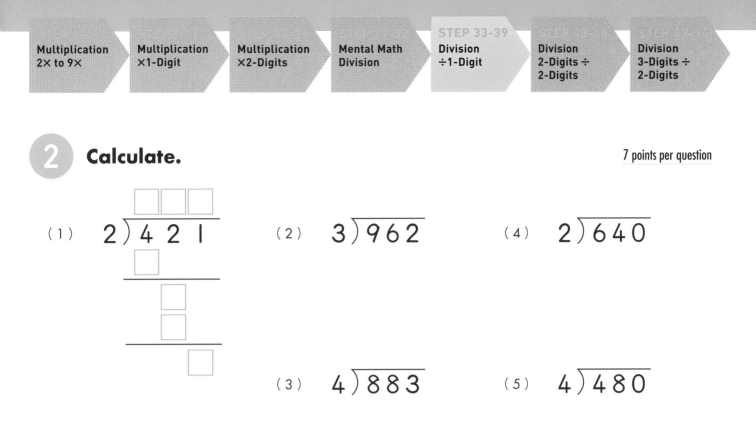

(1) 2)421

(2) 3)962

(4) 2)640

(3) 4)883

(5) 4)480

3 Calculate.

7 points per question

(1) 3)625

(2) 4)834

(4) 2)807

(3) 6)631

(5) 5)549

Review STEP 37

Calculate.

(1) 4)846

(2) 5)563

(3) 7)718

1 Calculate.

5 points per question

Example ● How to calculate 734 ÷ 5.

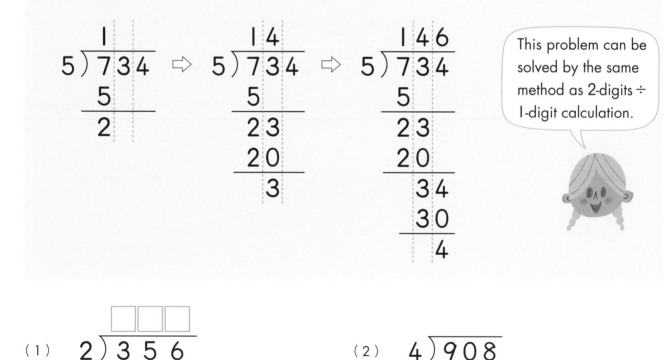

This problem can be solved by the same method as 2-digits ÷ 1-digit calculation.

(1) 2)356

(2) 4)908

2 Calculate.

10 points per question

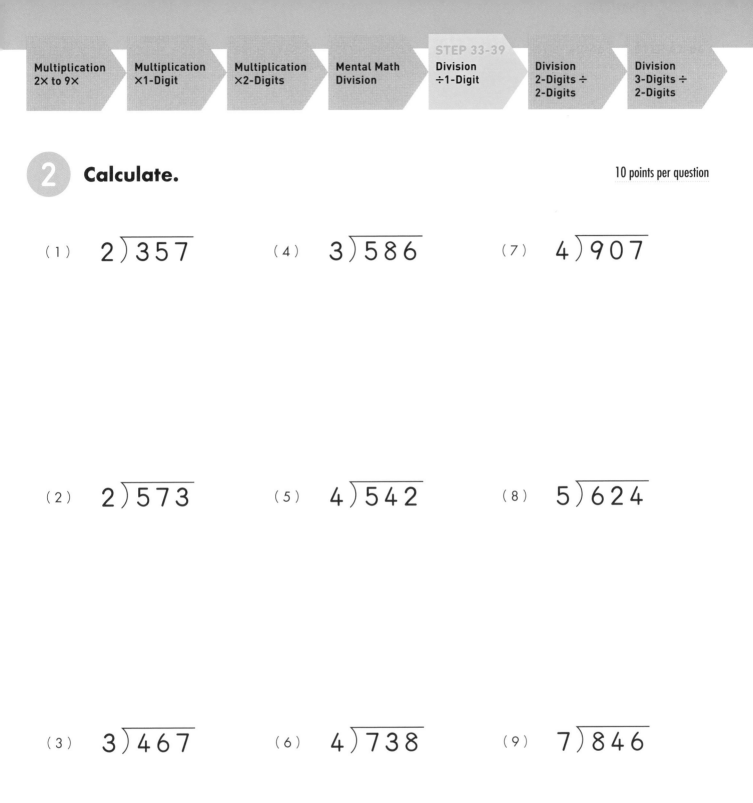

(1) $2\overline{)357}$

(4) $3\overline{)586}$

(7) $4\overline{)907}$

(2) $2\overline{)573}$

(5) $4\overline{)542}$

(8) $5\overline{)624}$

(3) $3\overline{)467}$

(6) $4\overline{)738}$

(9) $7\overline{)846}$

STEP **39**

Division ÷1-Digit

3-Digits÷1-Digit 3

Date   /   /

Score

/100

Review STEP 38

Calculate.

(1) 5)846

(2) 5)630

(3) 7)819

1 **Calculate.**

10 points per question

Example How to calculate 256÷4.

```
      6
  4)2 5 6
    2 4
      1
```
⇒
```
      6 4
  4)2 5 6
    2 4
      1 6
      1 6
        0
```

Here are division problems whose quotients are to be placed in the tens place.

(1) 5)426
```
      8 5
  5)4 2 6
    4 0
      2 6
      2 5
        1
```

(2) 3)253

(3) 4)345

2 **Calculate.**

8 points per question

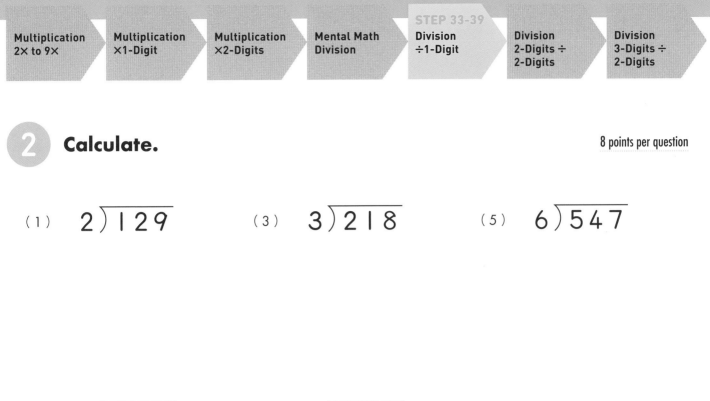

(1) $2\overline{)129}$

(3) $3\overline{)218}$

(5) $6\overline{)547}$

(2) $4\overline{)368}$

(4) $7\overline{)219}$

3 **Calculate.**

10 points per question

(1) $7\overline{)423}$

(4) $9\overline{)720}$

(7) $6\overline{)365}$

The number in the ones place of the quotient will be 0.

Review STEP **33** – STEP **36** **Calculate.** 4 points per question

(1) 2)96

(3) 5)67

(5) 8)93

(2) 6)67

(4) 8)89

Review STEP **37** **Calculate.** 5 points per question

(1) 2)623

(3) 5)526

(5) 4)884

(2) 3)685

(4) 3)967

(6) 2)680

Review **STEP 38** **Calculate.**

4 points per question

(1) $6\overline{)835}$

(3) $5\overline{)734}$

(5) $5\overline{)622}$

(2) $8\overline{)991}$

(4) $4\overline{)906}$

Review **STEP 39** **Calculate.**

5 points per question

(1) $4\overline{)228}$

(3) $6\overline{)462}$

(5) $9\overline{)360}$

(2) $5\overline{)365}$

(4) $7\overline{)364}$

(6) $3\overline{)245}$

2-Digits÷2-Digits 1

Review STEP 39

Calculate.

(1) $3\overline{)279}$

(2) $4\overline{)324}$

(3) $7\overline{)493}$

---

1 **Calculate.**

5 points per question

Example ● How to calculate $45 \div 21$.

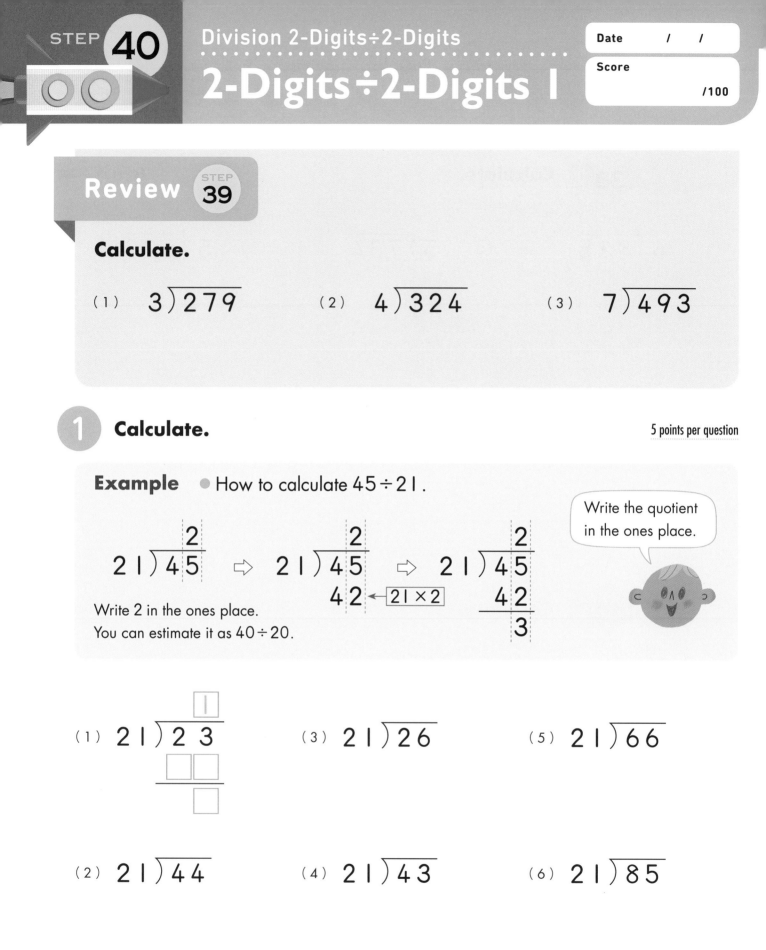

$$21\overline{)45}\ \ \ \ 2$$

Write 2 in the ones place.
You can estimate it as $40 \div 20$.

$$\Rightarrow\ \ 21\overline{)45}\ \ \ \ 2 \\ 42 \leftarrow \boxed{21 \times 2}$$

$$\Rightarrow\ \ 21\overline{)45}\ \ \ \ 2 \\ 42 \\ \ \ 3$$

Write the quotient in the ones place.

(1) $21\overline{)23}$

(3) $21\overline{)26}$

(5) $21\overline{)66}$

(2) $21\overline{)44}$

(4) $21\overline{)43}$

(6) $21\overline{)85}$

2 Calculate.

5 points per question

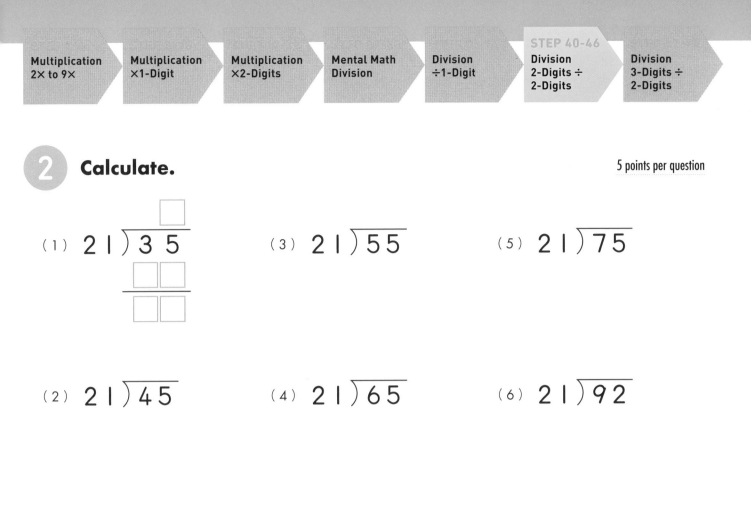

(1) 21)35

(3) 21)55

(5) 21)75

(2) 21)45

(4) 21)65

(6) 21)92

3 Calculate.

8 points per question

(1) 31)67

(3) 31)85

(5) 41)94

(2) 31)76

(4) 41)89

Review STEP 40

Calculate.

(1) $21\overline{)44}$ (2) $21\overline{)45}$ (3) $21\overline{)46}$

1 **Calculate.**

5 points per question

Example ● How to calculate $65 ÷ 31$.

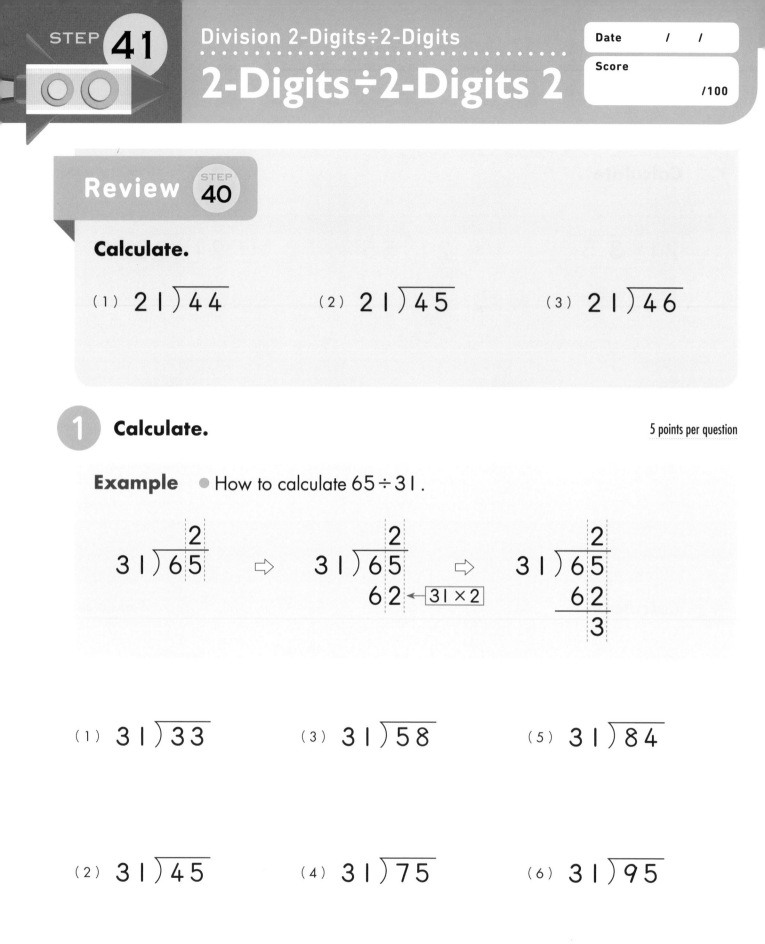

$$31\overline{)65} \quad \Rightarrow \quad 31\overline{)65} \atop 62 \leftarrow \boxed{31 \times 2} \quad \Rightarrow \quad 31\overline{)65} \atop 62 \atop 3$$

(1) $31\overline{)33}$ (3) $31\overline{)58}$ (5) $31\overline{)84}$

(2) $31\overline{)45}$ (4) $31\overline{)75}$ (6) $31\overline{)95}$

2 Calculate.

8 points per question

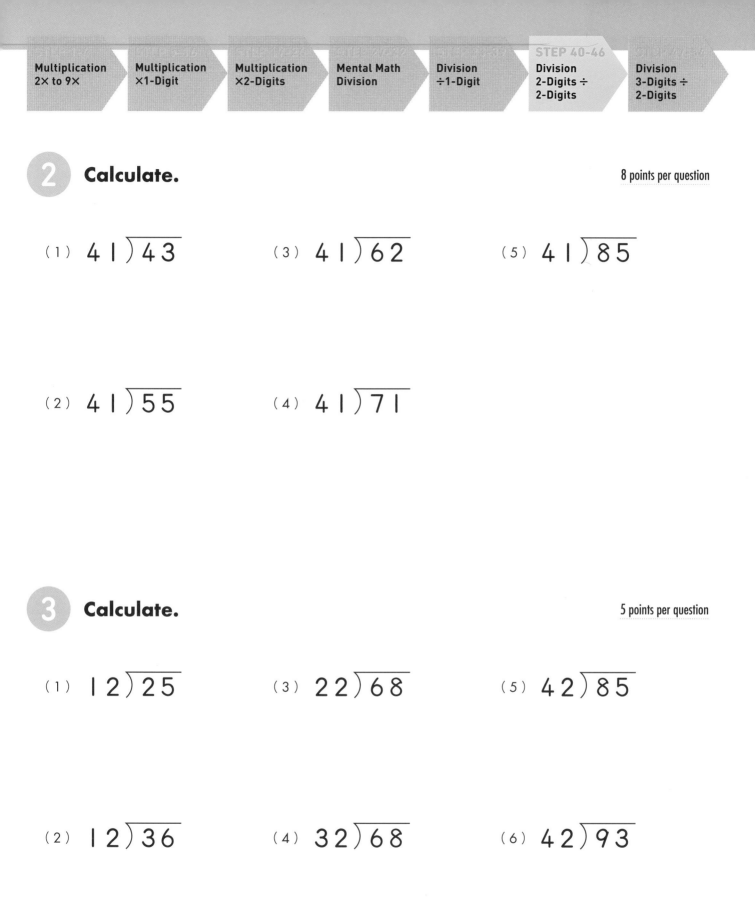

(1) 41)‾43‾

(2) 41)‾55‾

(3) 41)‾62‾

(4) 41)‾71‾

(5) 41)‾85‾

3 Calculate.

5 points per question

(1) 12)‾25‾

(2) 12)‾36‾

(3) 22)‾68‾

(4) 32)‾68‾

(5) 42)‾85‾

(6) 42)‾93‾

2-Digits÷2-Digits 3

Review STEP **41**

Calculate.

(1) 31)‾3‾4‾

(2) 41)‾8‾4‾

(3) 22)‾3‾8‾

1 **Calculate.**

9 points per question

Example ● How to calculate 61÷21.

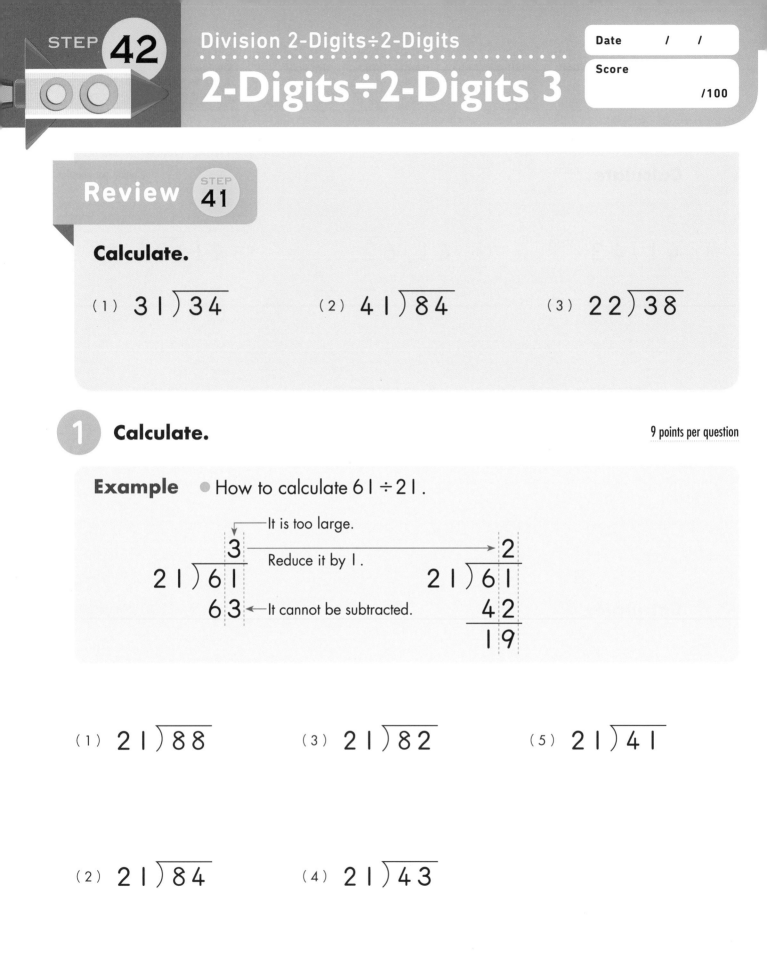

```
          3          It is too large.              2
       ┌──────   Reduce it by 1.    ──────▶    ┌──────
   21 )  6 1                             21 )  6 1
       6 3  ◀── It cannot be subtracted.      4 2
                                              ─────
                                              1 9
```

(1) 21)‾8‾8‾

(3) 21)‾8‾2‾

(5) 21)‾4‾1‾

(2) 21)‾8‾4‾

(4) 21)‾4‾3‾

2 **Calculate.**

5 points per question

(1) $21\overline{)70}$

(4) $31\overline{)65}$

(7) $21\overline{)64}$

(2) $21\overline{)80}$

(5) $31\overline{)62}$

(8) $21\overline{)60}$

(3) $21\overline{)90}$

(6) $31\overline{)60}$

3 **Calculate.**

5 points per question

(1) $22\overline{)66}$

(2) $22\overline{)65}$

(3) $22\overline{)86}$

2-Digits ÷ 2-Digits 4

Review STEP 42

Calculate.

(1) $21\overline{)54}$ (2) $21\overline{)61}$ (3) $21\overline{)81}$

1 Calculate.

5 points per question

Example ● How to calculate $66 \div 23$.

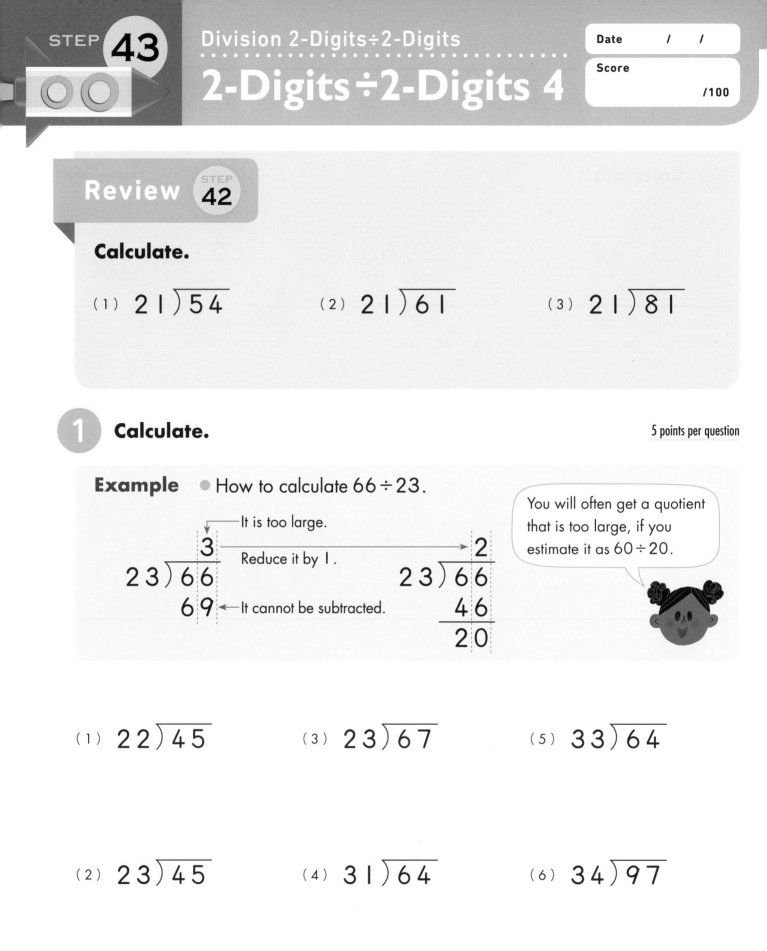

It is too large.

$$23\overline{)66} \quad \begin{array}{r} 3 \\ \end{array}$$
Reduce it by 1.
$$\begin{array}{r} 69 \end{array}$$ ← It cannot be subtracted.

$$\begin{array}{r} 2 \\ 23\overline{)66} \\ 46 \\ \hline 20 \end{array}$$

You will often get a quotient that is too large, if you estimate it as $60 \div 20$.

(1) $22\overline{)45}$ (3) $23\overline{)67}$ (5) $33\overline{)64}$

(2) $23\overline{)45}$ (4) $31\overline{)64}$ (6) $34\overline{)97}$

2 Calculate.

5 points per question

(1) $22 \overline{)42}$

(2) $22 \overline{)66}$

(3) $22 \overline{)65}$

(4) $43 \overline{)87}$

(5) $43 \overline{)85}$

(6) $43 \overline{)80}$

(7) $32 \overline{)66}$

(8) $32 \overline{)64}$

(9) $32 \overline{)61}$

(10) $33 \overline{)67}$

(11) $33 \overline{)64}$

(12) $34 \overline{)83}$

(13) $32 \overline{)96}$

(14) $33 \overline{)93}$

2-Digits÷2-Digits 5

Review STEP **43**

Calculate.

(1) 22)‾43‾

(2) 34)‾65‾

(3) 12)‾34‾

1 Calculate.

6 points per question

Example ● How to calculate 86÷25.

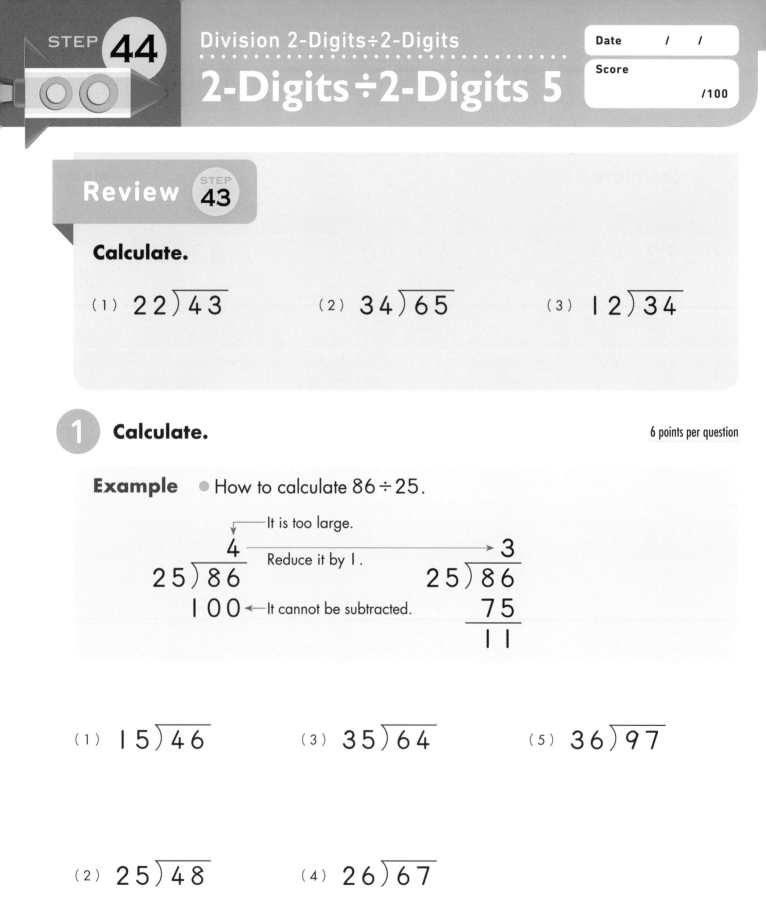

It is too large.

4

25)‾86‾

100 ← It cannot be subtracted.

Reduce it by 1.

3

25)‾86‾

75

11

(1) 15)‾46‾

(3) 35)‾64‾

(5) 36)‾97‾

(2) 25)‾48‾

(4) 26)‾67‾

2 Calculate.

5 points per question

(1) $16\overline{)38}$

(2) $25\overline{)65}$

(3) $26\overline{)89}$

(4) $35\overline{)94}$

(5) $36\overline{)69}$

(6) $45\overline{)88}$

(7) $46\overline{)89}$

(8) $25\overline{)98}$

(9) $26\overline{)73}$

(10) $36\overline{)88}$

(11) $35\overline{)98}$

(12) $15\overline{)36}$

(13) $35\overline{)71}$

(14) $36\overline{)71}$

2-Digits÷2-Digits 6

Date / /

Score /100

Review STEP 44

Calculate.

(1) $22\overline{)81}$ (2) $23\overline{)87}$ (3) $21\overline{)83}$

1 Calculate.

6 points per question

Example ● How to calculate $86 \div 28$.

$$\begin{array}{r} 4 \\ 28\overline{)86} \\ 112 \end{array}$$

4 ← It is too large.

Reduce it by 1. → 3

112 ← It cannot be subtracted.

$$\begin{array}{r} 3 \\ 28\overline{)86} \\ \underline{84} \\ 2 \end{array}$$

(1) $48\overline{)86}$ (3) $38\overline{)95}$ (5) $29\overline{)90}$

(2) $27\overline{)66}$ (4) $39\overline{)65}$

2 Calculate.

5 points per question

(1) $37\overline{)96}$

(2) $39\overline{)72}$

(3) $28\overline{)96}$

(4) $28\overline{)76}$

(5) $28\overline{)71}$

(6) $27\overline{)98}$

(7) $29\overline{)78}$

(8) $29\overline{)68}$

(9) $29\overline{)89}$

(10) $27\overline{)89}$

(11) $47\overline{)89}$

(12) $37\overline{)67}$

(13) $37\overline{)97}$

(14) $37\overline{)94}$

Review STEP 45

Calculate.

(1) $12\overline{)85}$ (2) $24\overline{)65}$ (3) $43\overline{)80}$

1   **Calculate.**

5 points per question

Example ● How to calculate $82 \div 12$.

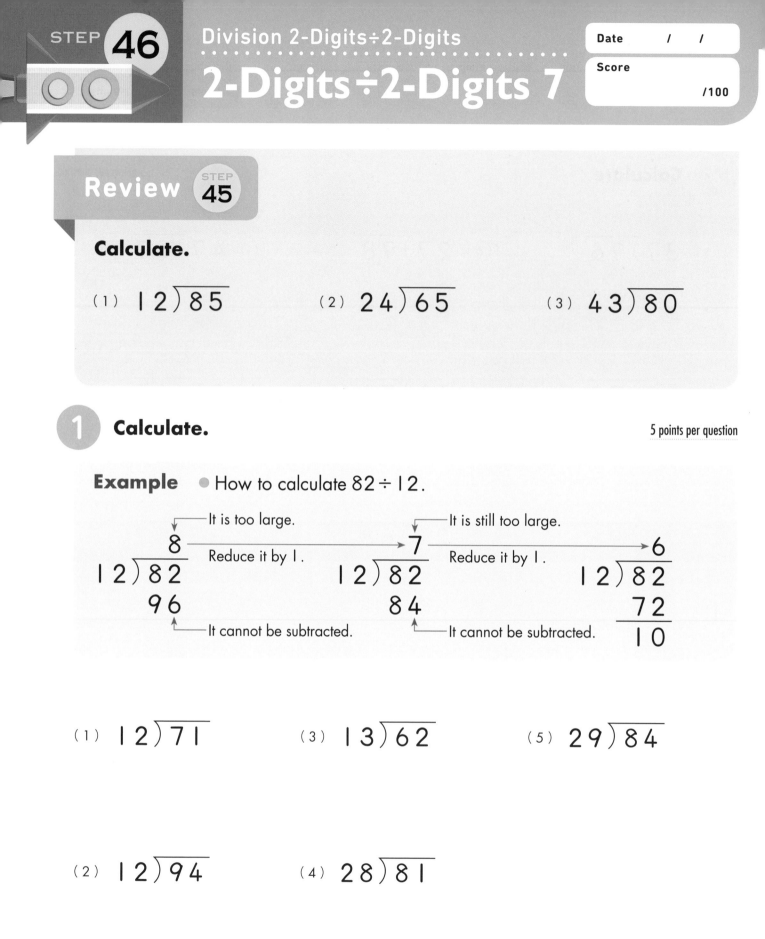

It is too large.

$$12\overline{)82} \quad \begin{array}{r} 8 \\ \hline 82 \\ 96 \end{array}$$

Reduce it by 1.

It cannot be subtracted.

It is still too large.

$$12\overline{)82} \quad \begin{array}{r} 7 \\ \hline 82 \\ 84 \end{array}$$

Reduce it by 1.

It cannot be subtracted.

$$12\overline{)82} \quad \begin{array}{r} 6 \\ \hline 82 \\ 72 \\ \hline 10 \end{array}$$

(1) $12\overline{)71}$ (3) $13\overline{)62}$ (5) $29\overline{)84}$

(2) $12\overline{)94}$ (4) $28\overline{)81}$

2 Calculate.

5 points per question

(1) $14\overline{)54}$

(2) $14\overline{)78}$

(3) $14\overline{)86}$

(4) $15\overline{)69}$

(5) $15\overline{)76}$

(6) $15\overline{)79}$

(7) $16\overline{)54}$

(8) $16\overline{)57}$

(9) $16\overline{)68}$

(10) $17\overline{)69}$

(11) $17\overline{)56}$

(12) $27\overline{)80}$

(13) $18\overline{)46}$

(14) $18\overline{)58}$

(15) $29\overline{)83}$

Division
2-Digits ÷ 2-Digits

Review STEP **40** **Calculate.**

4 points per question

(1) 21)24

(3) 21)45

(5) 21)84

(2) 21)68

(4) 21)36

Review STEP **41** STEP **42** **Calculate.**

5 points per question

(1) 31)96

(3) 42)45

(5) 41)57

(2) 41)87

(4) 42)94

(6) 31)67

Review STEP **43** – STEP **45**   **Calculate.**   4 points per question

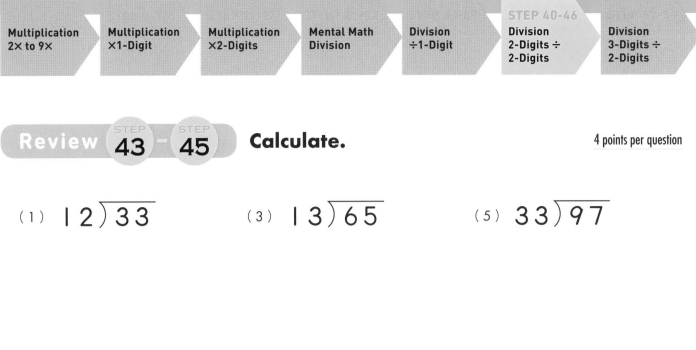

(1) $12\overline{)33}$

(2) $23\overline{)89}$

(3) $13\overline{)65}$

(4) $32\overline{)92}$

(5) $33\overline{)97}$

Review STEP **46**   **Calculate.**   5 points per question

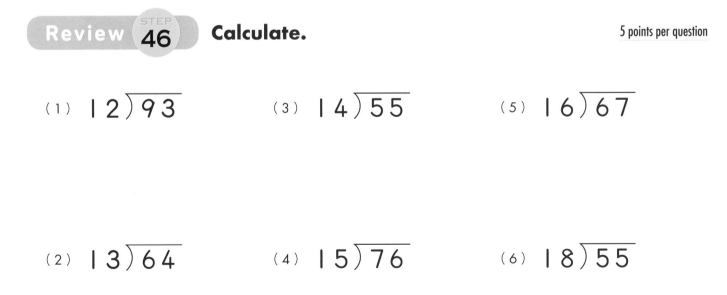

(1) $12\overline{)93}$

(2) $13\overline{)64}$

(3) $14\overline{)55}$

(4) $15\overline{)76}$

(5) $16\overline{)67}$

(6) $18\overline{)55}$

Review STEP 46

Calculate.

(1) 12)92

(2) 18)56

---

1 **Calculate.**

10 points per question

Example ● How to calculate 149÷21.

```
        7              7
21)149    ⇨    21)149
  147            147
                   2
```

It can be solved the same way as 2-digits÷2-digits calculation.

(1) 21)109

(3) 21)129

(2) 21)119

(4) 21)148

2 **Calculate.**

6 points per question

(1) $31\overline{)125}$

(3) $41\overline{)168}$

(2) $31\overline{)187}$

(4) $41\overline{)248}$

3 **Calculate.**

6 points per question

(1) $51\overline{)118}$

(4) $71\overline{)448}$

(2) $61\overline{)193}$

(5) $61\overline{)339}$

(3) $51\overline{)214}$

(6) $81\overline{)258}$

3-Digits÷2-Digits 2

Date / /

Score /100

Review STEP **47**

Calculate.

(1) $21\overline{)127}$

(2) $61\overline{)326}$

1 **Calculate.**

10 points per question

Example ● How to calculate 145÷43.

$$
\begin{array}{r}
3 \\
43\overline{)145} \\
129 \\
\hline
16
\end{array}
$$

Try calculating by making a broad estimation.

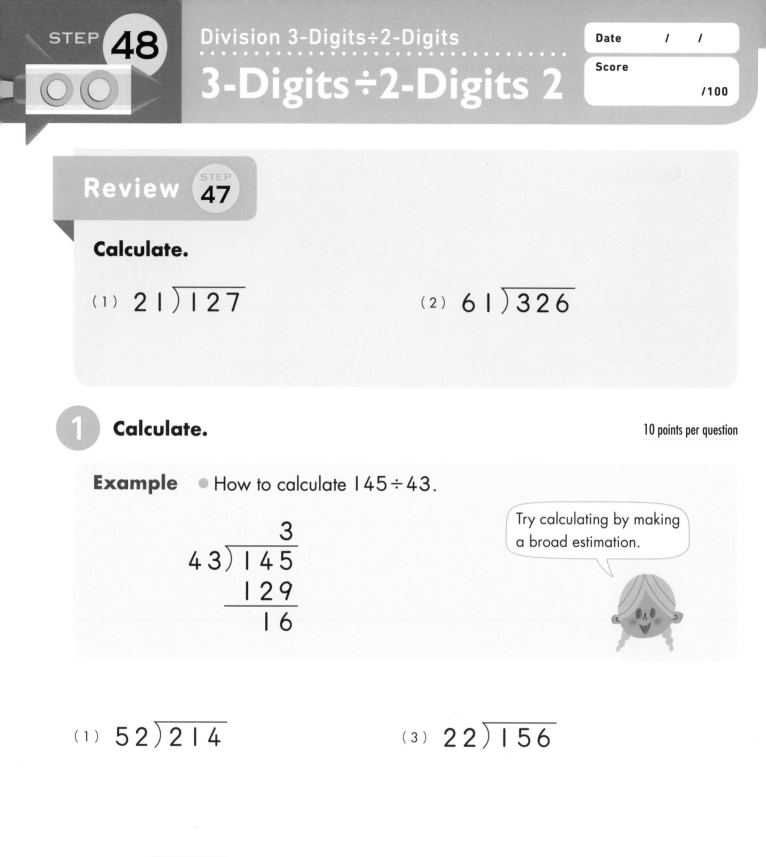

(1) $52\overline{)214}$

(3) $22\overline{)156}$

(2) $43\overline{)258}$

(4) $32\overline{)262}$

2 Calculate.

6 points per question

(1) $62\overline{)436}$

(2) $42\overline{)342}$

(3) $53\overline{)434}$

(4) $74\overline{)596}$

(5) $83\overline{)690}$

(6) $76\overline{)684}$

(7) $84\overline{)776}$

(8) $56\overline{)348}$

(9) $23\overline{)117}$

(10) $33\overline{)172}$

3-Digits÷2-Digits 3

Date / /

Score /100

Review STEP 48

Calculate.

(1) 22)179

(2) 86)364

1 Calculate.

10 points per question

Example ● How to calculate 188÷26.

(1) 45)245

(3) 54)427

(2) 24)153

(4) 36)291

© Kumon Publishing Co., Ltd.

2 Calculate.

6 points per question

(1) $45 \overline{)328}$

(6) $78 \overline{)601}$

(2) $57 \overline{)496}$

(7) $85 \overline{)653}$

(3) $64 \overline{)499}$

(8) $87 \overline{)734}$

(4) $69 \overline{)572}$

(9) $94 \overline{)656}$

(5) $76 \overline{)678}$

(10) $97 \overline{)747}$

STEP **50**

Division 3-Digits÷2-Digits

3-Digits ÷ 2-Digits 4

Date / /

Score

/100

Review STEP 49

Calculate.

(1) $32 \overline{)286}$

(2) $48 \overline{)340}$

1 Calculate.

20 points per question

Example ● How to calculate $234 \div 21$.

$$
\begin{array}{r}
1 \\
21{\overline{\smash{\big)}\,234}} \\
\underline{21} \\
2
\end{array}
\Rightarrow
\begin{array}{r}
11 \\
21{\overline{\smash{\big)}\,234}} \\
\underline{21} \\
24 \\
\underline{21} \\
3
\end{array}
$$

Write 1 (the answer for $23 \div 21$) in the tens place.

Move 4 in the ones place down.
$24 \div 21 = 1$ R 3

Here are division problems whose quotients are to be placed in the tens place and ones place.

(1) $21 \overline{)245}$

(2) $21 \overline{)445}$

2 Calculate.

10 points per question

(1) $31\overline{)674}$

(4) $37\overline{)885}$

(2) $43\overline{)924}$

(5) $23\overline{)969}$

(3) $42\overline{)893}$

(6) $32\overline{)874}$

Review STEP 50

Calculate.

(1) $39\overline{)842}$

(2) $22\overline{)781}$

1 Calculate.

10 points per question

Example ● How to calculate $682 \div 28$.

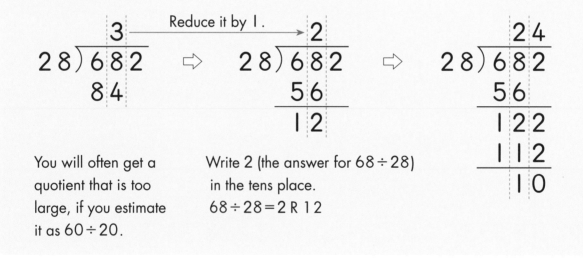

You will often get a quotient that is too large, if you estimate it as $60 \div 20$.

Write 2 (the answer for $68 \div 28$) in the tens place.
$68 \div 28 = 2 \text{ R } 12$

(1) $38\overline{)895}$

(2) $27\overline{)713}$

Multiplication
2× to 9×

Multiplication
×1-Digit

Multiplication
×2-Digits

Mental Math
Division

Division
÷1-Digit

Division
2-Digits ÷
2-Digits

STEP 47-54
Division
3-Digits ÷
2-Digits

2 Calculate.

10 points per question

(1) $32\overline{)769}$

(5) $35\overline{)941}$

(2) $27\overline{)975}$

(6) $38\overline{)998}$

(3) $26\overline{)941}$

(7) $23\overline{)864}$

(4) $26\overline{)936}$

(8) $34\overline{)974}$

STEP **52**

Division 3-Digits÷2-Digits
··
3-Digits÷2-Digits 6

Date / /

Score

/100

Review STEP 51

Calculate.

(1) $32\overline{)884}$

(2) $27\overline{)918}$

1 Calculate.

20 points

Example

● How to calculate 941÷23.

$$
\begin{array}{r}
40 \\
23\overline{)941} \\
92 \quad \leftarrow \boxed{23\times4} \\
\hline
21 \\
0 \quad \leftarrow \boxed{23\times0} \text{ You can omit it.} \\
\hline
21
\end{array}
$$

$32\overline{)654}$

2 Calculate.

10 points per question

(1) $14\overline{)572}$

(2) $24\overline{)724}$

(3) $14\overline{)852}$

(4) $24\overline{)975}$

(5) $26\overline{)784}$

(6) $29\overline{)897}$

(7) $32\overline{)973}$

(8) $46\overline{)926}$

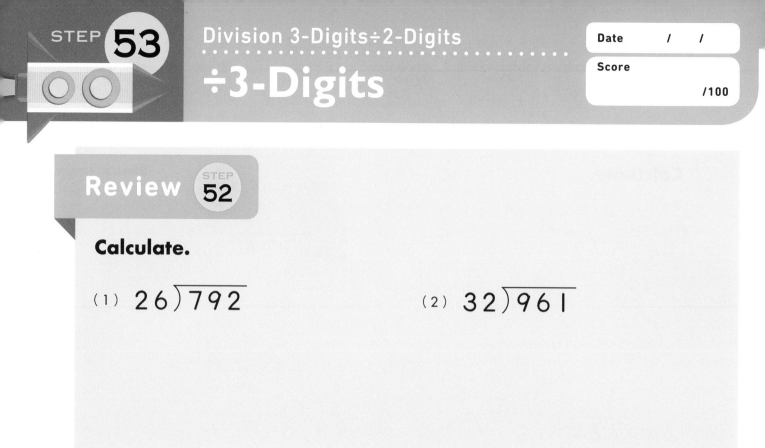

Review STEP **52**

Calculate.

(1) $26\overline{)792}$

(2) $32\overline{)961}$

 Calculate.

10 points per question

Example

● How to calculate $501 ÷ 123$.

$$\begin{array}{r} 4 \\ 123\overline{)501} \\ \underline{492} \leftarrow \boxed{123 \times 4} \\ 9 \end{array}$$

(2) $237\overline{)948}$

(1) $164\overline{)997}$

(3) $306\overline{)957}$

2 Calculate.

14 points per question

Example

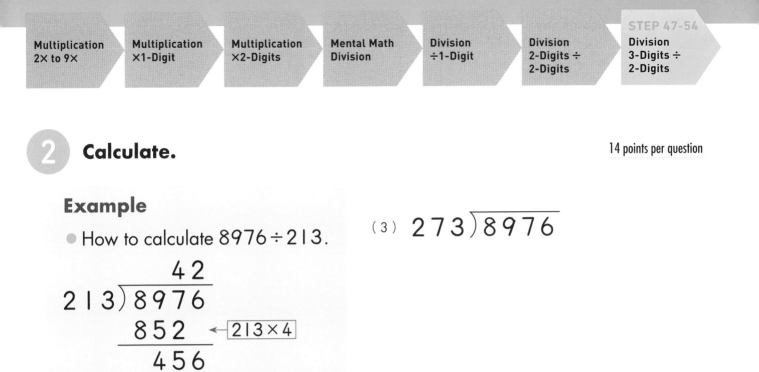

● How to calculate 8976÷213.

```
            42
    213)8976
        852    ←  213×4
        456
        426    ←  213×2
         30
```

(3) 273)8976

(1) 223)8976

(4) 283)8976

(2) 253)8976

(5) 293)8976

Review STEP 53

Calculate.

(1) $342\overline{)819}$

(2) $263\overline{)8976}$

1 Calculate.

16 points

Example

● How to calculate $4876 \div 21$.

$21\overline{)6789}$

```
          2 3 2
    21 ) 4 8 7 6
          4 2      ← 21 × 2
          6 7
          6 3      ← 21 × 3
          4 6
          4 2      ← 21 × 2
            4
```

2 Calculate.

14 points per question

(1) $21\overline{)5678}$

(4) $32\overline{)6421}$

(2) $31\overline{)8902}$

(5) $32\overline{)4000}$

(3) $31\overline{)6799}$

(6) $42\overline{)6789}$

STEP 47 - STEP 54 TEST

Division 3-Digits÷2-Digits

Date / /

Score /100

Review STEP 47 STEP 48 **Calculate.** 5 points per question

(1) $21\overline{)149}$

(3) $25\overline{)127}$

(2) $32\overline{)256}$

(4) $71\overline{)358}$

Review STEP 49 STEP 50 **Calculate.** 6 points per question

(1) $42\overline{)324}$

(3) $23\overline{)284}$

(2) $34\overline{)271}$

(4) $19\overline{)412}$

Review STEP **51** STEP **52**   **Calculate.**   6 points per question

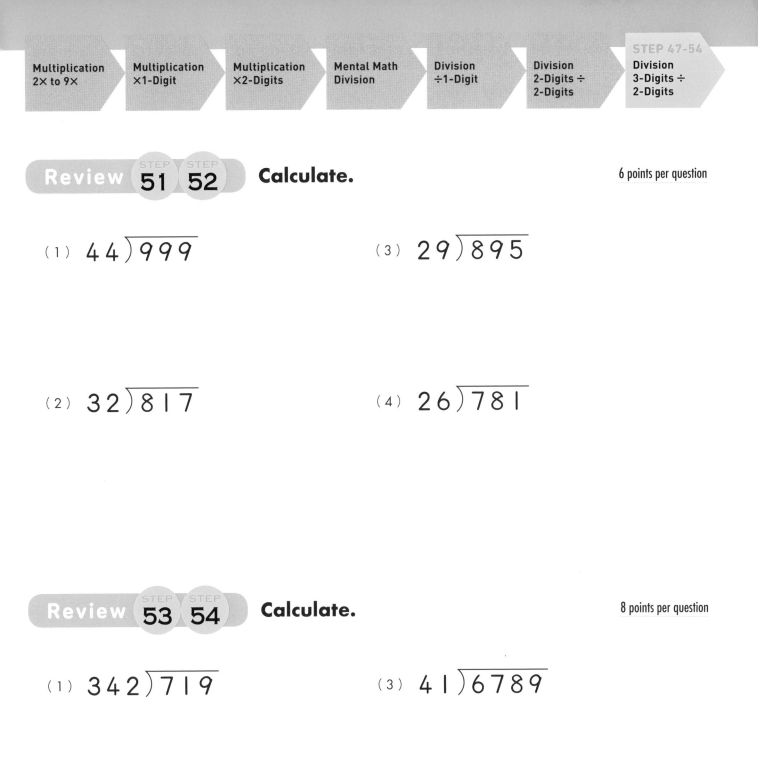

(1) 44)999

(3) 29)895

(2) 32)817

(4) 26)781

Review STEP **53** STEP **54**   **Calculate.**   8 points per question

(1) 342)719

(3) 41)6789

(2) 362)7986

(4) 44)8902

Math Boosters

Grades 2-4 Multiplication & Division

Answer Key

STEP 1 (P.4 · 5)

■ Fill in the missing numbers in the boxes below.

(1) 10, 12   (2) 18, 20   (3) 15, 18

① · 2×
(1) 2
(2) 4
(3) 6
(4) 8
(5) 10
(6) 12
(7) 14
(8) 16
(9) 18

· 3×
(1) 3
(2) 6
(3) 9
(4) 12
(5) 15
(6) 18
(7) 21
(8) 24
(9) 27

②
(1) 2   (4) 4   (7) 14
(2) 10   (5) 8   (8) 18
(3) 6   (6) 16

③
(1) 3   (4) 6   (7) 15
(2) 12   (5) 21   (8) 27
(3) 18   (6) 9   (9) 24

④
(1) 3   (6) 9
(2) 4   (7) 8
(3) 5   (8) 7
(4) 1   (9) 6
(5) 2   (10) 5

STEP 2 (P.6 · 7)

■ Fill in the missing numbers in the boxes below.

(1) 20, 24   (2) 32, 36   (3) 25, 30

---

① · 4×
(1) 4
(2) 8
(3) 12
(4) 16
(5) 20
(6) 24
(7) 28
(8) 32
(9) 36

· 5×
(1) 5
(2) 10
(3) 15
(4) 20
(5) 25
(6) 30
(7) 35
(8) 40
(9) 45

②
(1) 4   (4) 8   (7) 28
(2) 20   (5) 16   (8) 36
(3) 12   (6) 32

③
(1) 5   (4) 10   (7) 25
(2) 20   (5) 35   (8) 45
(3) 30   (6) 15   (9) 40

④
(1) 3   (6) 9
(2) 2   (7) 8
(3) 5   (8) 7
(4) 1   (9) 6
(5) 2   (10) 5

STEP 3 (P.8 · 9)

■ Fill in the missing numbers in the boxes below.

(1) 30, 36   (2) 54, 60   (3) 35, 42

① · 6×
(1) 6
(2) 12
(3) 18
(4) 24
(5) 30
(6) 36
(7) 42
(8) 48
(9) 54

· 7×
(1) 7
(2) 14
(3) 21
(4) 28
(5) 35
(6) 42
(7) 49
(8) 56
(9) 63

②
(1) 6
(2) 30
(3) 18
(4) 12
(5) 24
(6) 48
(7) 42
(8) 54
(9) 36

③
(1) 7
(2) 28
(3) 42
(4) 14
(5) 49
(6) 56
(7) 35
(8) 63

④
(1) 3
(2) 2
(3) 5
(4) 1
(5) 2
(6) 9
(7) 8
(8) 7
(9) 6
(10) 5

STEP 4

(P.10 · 11)

■ Fill in the missing numbers in the boxes below.

(1) 40, 48　　(2) 72, 80　　(3) 45, 54

❶

· 8×
(1) 8
(2) 16
(3) 24
(4) 32
(5) 40
(6) 48
(7) 56
(8) 64
(9) 72

· 9×
(1) 9
(2) 18
(3) 27
(4) 36
(5) 45
(6) 54
(7) 63
(8) 72
(9) 81

❷
(1) 8
(2) 40
(3) 24
(4) 16
(5) 32
(6) 64
(7) 56
(8) 72
(9) 48

❸
(1) 9
(2) 36
(3) 54
(4) 18
(5) 63
(6) 27
(7) 45
(8) 81

④
(1) 3
(2) 2
(3) 5
(4) 1
(5) 2
(6) 9
(7) 8
(8) 7
(9) 6
(10) 5

TEST

(P.12 · 13)

Review of Step 1,2
(1) 18
(2) 24
(3) 28
(4) 30
(5) 10
(6) 12
(7) 12
(8) 10
(9) 21
(10) 20
(11) 9
(12) 45
(13) 16
(14) 24
(15) 35
(16) 32
(17) 27
(18) 12
(19) 36
(20) 15
(21) 6
(22) 16
(23) 14
(24) 40
(25) 18

Review of Step 3,4
(1) 56
(2) 35
(3) 63
(4) 48
(5) 48
(6) 30
(7) 16
(8) 24
(9) 72
(10) 14
(11) 42
(12) 64
(13) 49
(14) 63
(15) 54
(16) 40
(17) 18
(18) 72
(19) 42
(20) 36
(21) 21
(22) 24
(23) 56
(24) 54
(25) 81

STEP 5

(P.14 · 15)

■ **Review of Step 1,2**
(1) 12
(2) 14
(3) 36
(4) 6
(5) 8
(6) 4

❶
(1) 24
(2) 28
(3) 44
(4) 48
(5) 64
(6) 86

2 (1) 33　　(3) 63　　(5) 93
　　(2) 66　　(4) 69　　(6) 96

3 (1) 48　　(3) 84
　　(2) 44　　(4) 88

STEP 6　　　　　　　　　　(P.16 · 17)

■Review of Step 5
　　(1) 24　　(2) 69　　(3) 44

1 (1) 36　　(3) 38　　(5) 48
　　(2) 34　　(4) 42　　(6) 54

2 (1) 52　　(4) 58　　(7) 84
　　(2) 56　　(5) 75　　(8) 78
　　(3) 54　　(6) 72

3 (1) 74　　(3) 76　　(5) 96
　　(2) 72　　(4) 98

STEP 7　　　　　　　　　　(P.18 · 19)

■Review of Step 6
　　(1) 32　　(2) 51　　(3) 78

1 (1) 52　　(3) 64　　(5) 72
　　(2) 56　　(4) 68

2 (1) 65　　(6) 84　　(11) 96
　　(2) 75　　(7) 95　　(12) 84
　　(3) 85　　(8) 76　　(13) 91
　　(4) 72　　(9) 96　　(14) 98
　　(5) 78　　(10) 92　　(15) 96

STEP 8　　　　　　　　　　(P.20 · 21)

■Review of Step 7
　　(1) 76　　(2) 96　　(3) 92

1 (1) 123　　(3) 156　　(5) 159
　　(2) 129　　(4) 153

2 (1) 126　　(6) 213　　(11) 142
　　(2) 189　　(7) 216　　(12) 148
　　(3) 128　　(8) 183　　(13) 246
　　(4) 186　　(9) 168　　(14) 186
　　(5) 144　　(10) 219　　(15) 279

STEP 9　　　　　　　　　　(P.22 · 23)

■Review of Step 8
　　(1) 124　　(2) 123　　(3) 104

1 (1) 112　　(3) 114　　(5) 172
　　(2) 136　　(4) 158

2 (1) 171　　(5) 231　　(9) 252
　　(2) 174　　(6) 195　　(10) 237
　　(3) 168　　(7) 234　　(11) 264
　　(4) 228　　(8) 258　　(12) 261

3 (1) 204　　(2) 201　　(3) 207

STEP 10　　　　　　　　　　(P.24 · 25)

■Review of Step 9
　　(1) 134　　(2) 204　　(3) 112

1 (1) 144　　(3) 152　　(5) 125
　　(2) 148　　(4) 115

2 (1) 240　　(6) 252　　(11) 344
　　(2) 224　　(7) 192　　(12) 430
　　(3) 180　　(8) 115　　(13) 435
　　(4) 104　　(9) 160　　(14) 304
　　(5) 116　　(10) 395　　(15) 420

STEP 11

(P.26 · 27)

■ Review of Step 10

(1) 148   (2) 215   (3) 108

1 (1) 138   (3) 390   (5) 686
(2) 282   (4) 154

2 (1) 102   (6) 408   (11) 616
(2) 133   (7) 602   (12) 623
(3) 343   (8) 522   (13) 174
(4) 413   (9) 511   (14) 301
(5) 322   (10) 539   (15) 504

STEP 12

(P.28 · 29)

■ Review of Step 11

(1) 112   (2) 161   (3) 228

1 (1) 256   (3) 432   (5) 784
(2) 477   (4) 738

2 (1) 280   (6) 261   (11) 536
(2) 207   (7) 392   (12) 522
(3) 136   (8) 405   (13) 304
(4) 384   (9) 520   (14) 306
(5) 333   (10) 711   (15) 504

TEST

(P.30 · 31)

Review of Step 5

(1) 48   (3) 48   (5) 88
(2) 69   (4) 96   (6) 84

Review of Step 6-8

(1) 72   (3) 52   (5) 186
(2) 75   (4) 126

Review of Step 9

(1) 112   (2) 130   (3) 234   (4) 267

Review of Step 10

(1) 185   (3) 232   (5) 230
(2) 170   (4) 276   (6) 348

Review of Step 11

(1) 175   (3) 448   (5) 364
(2) 228   (4) 468

Review of Step 12

(1) 416   (2) 621   (3) 288   (4) 387

STEP 13

(P.32 · 33)

■ Review of Step 12

(1) 432   (2) 423   (3) 312

1 (1) 363   (3) 366   (5) 399
(2) 242   (4) 488

2 (1) 636   (5) 468   (9) 684
(2) 663   (6) 699   (10) 804
(3) 444   (7) 628   (11) 609
(4) 884   (8) 969   (12) 909

STEP 14

(P.34 · 35)

■ Review of Step 13

(1) 264   (2) 888   (3) 606

1 (1) 852   (3) 675   (5) 681
(2) 632   (4) 452   (6) 978

2 (1) 572   (6) 512   (11) 858
(2) 462   (7) 734   (12) 966
(3) 334   (8) 978   (13) 992
(4) 735   (9) 665   (14) 954
(5) 992   (10) 520

STEP 15 (P.36 · 37)

■ Review of Step 14
(1) 852　　(2) 468　　(3) 624

1 (1) 1248　　(2) 1536　　(3) 2466

2 (1) 1305　　(2) 1968　　(3) 1926

3 (1) 1552　　(4) 3498　　(7) 3748
(2) 1985　　(5) 3470　　(8) 7992
(3) 2868　　(6) 6732

4 (1) 1016　　(3) 2088　　(5) 6125
(2) 1029　　(4) 3353

STEP 16 (P.38 · 39)

■ Review of Step 15
(1) 2265　　(2) 2316　　(3) 2082

1 (1) 8642　　(3) 4936　　(5) 9544
(2) 9087　　(4) 7185　　(6) 9765

2 (1) 17284　　(6) 25608　　(11) 43416
(2) 10296　　(7) 12670　　(12) 47915
(3) 22456　　(8) 15140　　(13) 59088
(4) 36001　　(9) 27918　　(14) 89991
(5) 39072　　(10) 37632

TEST (P.40 · 41)

Review of Step 13
(1) 396　　(3) 406　　(5) 969
(2) 286　　(4) 884

Review of Step 14
(1) 357　　(3) 948　　(5) 620
(2) 892　　(4) 972

Review of Step 15,16
(1) 1575　　(6) 2724　　(11) 8878
(2) 1264　　(7) 6013　　(12) 47684
(3) 1449　　(8) 8631　　(13) 8073
(4) 2990　　(9) 7392　　(14) 59562
(5) 5080　　(10) 6425　　(15) 74871

STEP 17 (P.42 · 43)

■ Review of Step 16
(1) 4926　　(2) 28044

1 (1) 390　　(3) 720　　(5) 900
(2) 480　　(4) 960

2 (1) 1400　　(5) 1050　　(9) 3480
(2) 1140　　(6) 2600　　(10) 1200
(3) 1840　　(7) 2450
(4) 2160　　(8) 5920

STEP 18 (P.44 · 45)

■ Review of Step 17
(1) 280　　(2) 2120　　(3) 1800

1 (1)
$$\begin{array}{r} 32 \\ \times\ 12 \\ \hline 6\boxed{4} \\ 3\boxed{2} \\ \hline 3\boxed{8}\boxed{4} \end{array}$$
(2) 352　　(3) 286

2 (1) 288　　(5) 882　　(9) 989
(2) 384　　(6) 837　　(10) 742
(3) 264　　(7) 598　　(11) 936
(4) 492　　(8) 966

STEP 19 (P.46 · 47)

■ Review of Step 18
(1) 264　　(2) 624　　(3) 448

① (1)
```
      2 7
  ×   1 3
  ─────────
    [8][1]
  [2][7]
  ─────────
  [3][5][1]
```
(2) 552

(3)
```
      4 2
  ×   1 7
  ─────────
  [2][9][4]
  [4][2]
  ─────────
  [7][1][4]
```

② (1) 506   (5) 512   (9) 608

(2) 810   (6) 858   (10) 952

(3) 504   (7) 644   (11) 900

(4) 416   (8) 800

STEP 20

■ **Review of Step 19**

(1) 364   (2) 720   (3) 621

① (1)
```
      4 2
  ×   3 3
  ─────────
  [1][2][6]
  [1][2][6]
  ─────────
  [1][3][8][6]
```
(2) 2244   (3) 3294

② (1) 4428   (5) 5336   (9) 6794

(2) 4554   (6) 5376   (10) 7656

(3) 4592   (7) 6525   (11) 8722

(4) 5330   (8) 6474

STEP 21
(P.50 · 51)

■ **Review of Step 20**

(1) 2244   (2) 4428   (3) 6468

① (1)
```
      4 7
  ×   2 3
  ─────────
  [1][4][1]
  [9][4]
  ─────────
  [1][0][8][1]
```
(2) 1175   (3) 1269

② (1) 1152   (5) 1161   (9) 1092

(2) 1292   (6) 1184   (10) 1064

(3) 1088   (7) 1176   (11) 1064

(4) 1092   (8) 1260

STEP 22
(P.52 · 53)

■ **Review of Step 21**

(1) 1092   (2) 1161   (3) 1092

① (1) 2072   (2) 2268   (3) 2457

② (1) 3136   (5) 7482   (9) 6586

(2) 3038   (6) 5607   (10) 8428

(3) 4002   (7) 4161   (11) 9312

(4) 5005   (8) 4218

STEP 23
(P.54 · 55)

■ **Review of Step 22**

(1) 2208   (2) 3136   (3) 4002

① (1)
```
      4 8
  ×   2 5
  ─────────
  [2][4][0]
  [9][6]
  ─────────
  [1][2][0][0]
```
(2) 1036   (3) 1104

② (1) 1105   (5) 1248   (9) 1026

(2) 1014   (6) 1044   (10) 1007

(3) 1222   (7) 1309   (11) 1128

(4) 1032   (8) 1206

STEP 24
(P.56 · 57)

■ **Review of Step 23**

(1) 1120   (2) 1254   (3) 1247

① (1)
```
      2 1 2
  ×     1 4
  ─────────
  [8][4][8]
  [2][1][2]
  ─────────
  [2][9][6][8]
```
(2) 9504

② (1) 4173 (5) 6048
(2) 7062 (6) 8650
(3) 7176 (7) 9440
(4) 6000 (8) 9246

STEP 25

■ Review of Step 24

(1) 7848 (2) 9440

① (1)
```
    426
  ×  48
  3408
 1704
 20448
```
(2)
```
    417
  ×  31
   417
  1251
  12927
```

② (1)
```
    332
  ×  31
    332
   996
  10292
```
(5)
```
    908
  ×  71
    908
  6356
  64468
```

(2)
```
    406
  ×  26
   2436
   812
  10556
```
(6)
```
    432
  ×  85
   2160
  3456
  36720
```

(3)
```
    324
  ×  53
    972
  1620
  17172
```
(7)
```
    134
  ×  85
    670
  1072
  11390
```

(4)
```
    417
  ×  32
    834
  1251
  13344
```
(8)
```
    268
  ×  45
   1340
  1072
  12060
```

STEP 26

(P.60 · 61)

■ Review of Step 25

(1)
```
    342
  ×  32
    684
  1026
  10944
```
(2)
```
    173
  ×  62
    346
  1038
  10726
```

① (1)
```
    324
  ×  42
    648
  1296
  13608
```
(2)
```
    314
  ×  68
   2512
  1884
  21352
```

② (1)
```
    324
  ×  33
    972
   972
  10692
```
(5)
```
    282
  ×  37
   1974
   846
  10434
```

(2)
```
    432
  ×  24
   1728
   864
  10368
```
(6)
```
    432
  ×  96
   2592
  3888
  41472
```

(3)
```
    280
  ×  37
   1960
   840
  10360
```
(7)
```
    697
  ×  78
   5576
  4879
  54366
```

(4)
```
    406
  ×  37
   2842
  1218
  15022
```
(8)
```
    326
  ×  33
    978
   978
  10758
```

Review of Step 17-23

(1)
```
    37
  × 22
    74
   74
   814
```

(5)
```
    33
  × 35
   165
    99
  1155
```

(9)
```
    87
  × 98
   696
   783
  8526
```

(2)
```
    41
  × 41
    41
   164
  1681
```

(6)
```
    46
  × 38
   368
   138
  1748
```

(10)
```
    97
  × 99
   873
   873
  9603
```

(3)
```
    45
  × 73
   135
   315
  3285
```

(7)
```
    37
  × 45
   185
   148
  1665
```

(4)
```
    42
  × 27
   294
    84
  1134
```

(8)
```
    64
  × 97
   448
   576
  6208
```

Review of Step 24-26

(1)
```
    321
  ×  21
    321
    642
   6741
```

(5)
```
    432
  ×  98
   3456
   3888
  42336
```

(2)
```
    534
  ×  58
   4272
   2670
  30972
```

(6)
```
    406
  ×  47
   2842
   1624
  19082
```

(3)
```
    806
  ×  61
    806
   4836
  49166
```

(7)
```
    887
  ×  98
   7096
   7983
  86926
```

(4)
```
    406
  ×  28
   3248
    812
  11368
```

(8)
```
    998
  ×  97
   6986
   8982
  96806
```

■ Review of Step 1

(1) 16　(3) 12　(5) 12
(2) 14　(4) 10　(6) 15

1
(1) $2 \times \boxed{2} = 4 \longrightarrow 4 \div 2 = \boxed{2}$
(2) $2 \times \boxed{3} = 6 \longrightarrow 6 \div 2 = \boxed{3}$
(3) $3 \times \boxed{2} = 6 \longrightarrow 6 \div 3 = \boxed{2}$
(4) $3 \times \boxed{3} = 9 \longrightarrow 9 \div 3 = \boxed{3}$

2
(1) 4　(5) 8　(9) 6
(2) 5　(6) 7　(10) 9
(3) 6　(7) 4　(11) 8
(4) 9　(8) 5　(12) 7

3
(1) 9　(4) 7　(7) 1
(2) 8　(5) 8　(8) 3
(3) 7　(6) 9　(9) 0

■ Review of Step 2,3

(1) 20　(3) 42　(5) 40
(2) 24　(4) 35　(6) 54

1
(1) $4 \times \boxed{2} = 8 \longrightarrow 8 \div 4 = \boxed{2}$
(2) $5 \times \boxed{3} = 15 \longrightarrow 15 \div 5 = \boxed{3}$
(3) $6 \times \boxed{2} = 12 \longrightarrow 12 \div 6 = \boxed{2}$
(4) $6 \times \boxed{3} = 18 \longrightarrow 18 \div 6 = \boxed{3}$

2
(1) 4　(7) 4　(13) 4
(2) 5　(8) 5　(14) 5
(3) 6　(9) 6　(15) 6
(4) 9　(10) 9　(16) 9
(5) 8　(11) 8　(17) 8
(6) 7　(12) 7　(18) 7

3
(1) 1　(3) 4
(2) 1　(4) 0

STEP ㉙

(P.68 · 69)

■ Review of Step 3,4

(1) 63   (3) 45   (5) 64
(2) 56   (4) 48   (6) 81

1
(1) 7 × ③ = 21 ⟶ 21 ÷ 7 = ③
(2) 8 × ③ = 24 ⟶ 24 ÷ 8 = ③
(3) 9 × ② = 18 ⟶ 18 ÷ 9 = ②
(4) 9 × ③ = 27 ⟶ 27 ÷ 9 = ③

2

(1) 4	(7) 4	(13) 4
(2) 5	(8) 5	(14) 5
(3) 6	(9) 6	(15) 6
(4) 9	(10) 9	(16) 9
(5) 8	(11) 8	(17) 8
(6) 7	(12) 7	(18) 7

3
(1) 1   (3) 1
(2) 8   (4) 0

STEP ㉚

(P.70 · 71)

■ Review of Step 27

(1) 5   (3) 7   (5) 5
(2) 8   (4) 6   (6) 9

1
(1) 2   (4) 2
(2) 2 R 1   (5) 2 R 1
(3) 3 R 1   (6) 3 R 1

2
(1) 4   (5) 9 R 1
(2) 4 R 1   (6) 8 R 1
(3) 6 R 1   (7) 5 R 1
(4) 7 R 1   (8) 9

3
(1) 4 R 1   (5) 6 R 1
(2) 4 R 2   (6) 6 R 2
(3) 5 R 2   (7) 9 R 1
(4) 5 R 1   (8) 8 R 2

STEP ㉛

(P.72 · 73)

■ Review of Step 28

(1) 2   (3) 6   (5) 5
(2) 6   (4) 2   (6) 9

1
(1) 2   (5) 2 R 2
(2) 2 R 1   (6) 2 R 3
(3) 2 R 2   (7) 2 R 4
(4) 2 R 3   (8) 3

2
(1) 4 R 2   (4) 5 R 3
(2) 4 R 1   (5) 9 R 1
(3) 5 R 2   (6) 7 R 2

3
(1) 6 R 3   (3) 9 R 1
(2) 5 R 2   (4) 8 R 4

4
(1) 4 R 2   (4) 8 R 2
(2) 5 R 2   (5) 8 R 4
(3) 7 R 3   (6) 9 R 3

STEP ㉜

(P.74 · 75)

■ Review of Step 29

(1) 5   (3) 7   (5) 5
(2) 6   (4) 4   (6) 7

1
(1) 2 R 6   (5) 3 R 2
(2) 3   (6) 3 R 1
(3) 3 R 1   (7) 3
(4) 3 R 2   (8) 2 R 7

2
(1) 4 R 3   (3) 7 R 3
(2) 5 R 6   (4) 9 R 2

3
(1) 4 R 4   (4) 7 R 3
(2) 5 R 3   (5) 8 R 1
(3) 7 R 1   (6) 9 R 3

4 (1) 4 R 2   (4) 6 R 3
(2) 4 R 1   (5) 8 R 3
(3) 5 R 3   (6) 9 R 2

TEST

(P.76 · 77)

Review of Step 30
(1) 7 R 1   (5) 8 R 1
(2) 6 R 1   (6) 8 R 1
(3) 3 R 1   (7) 4 R 1
(4) 4   (8) 7 R 2

Review of Step 31
(1) 4 R 1   (4) 7
(2) 4 R 3   (5) 9 R 3
(3) 7 R 2   (6) 7 R 3

Review of Step 32
(1) 4 R 3   (8) 8 R 1
(2) 4 R 4   (9) 8
(3) 4 R 2   (10) 9 R 4
(4) 7 R 4   (11) 9 R 3
(5) 6   (12) 9 R 2
(6) 6 R 4   (13) 9 R 4
(7) 9   (14) 9 R 5

STEP 33

(P.78 · 79)

■ Review of Step 32
(1) 2 R 2   (4) 5 R 6
(2) 7 R 1   (5) 8 R 1
(3) 8 R 3   (6) 7 R 2

1 (1)
```
     8
  4)3 2
    3 2
      0
```
(2) 4
(3) 4 R 1

2 (1) 7   (4) 4   (7) 9
(2) 7   (5) 6   (8) 5
(3) 6   (6) 4

3 (1) 6 R 2   (4) 8 R 3
(2) 7 R 3   (5) 8 R 6
(3) 4 R 1

STEP 34

(P.80 · 81)

■ Review of Step 33
(1) 8 R 1   (3) 4 R 4
(2) 5 R 2

1 (1) 12   (3) 15 R 1
(2) 12 R 1

2 (1) 21 R 1   (4) 39 R 1
(2) 28 R 1   (5) 38 R 1
(3) 15 R 2   (6) 42 R 1

3 (1) 17 R 2   (4) 24
(2) 21 R 2   (5) 29 R 1
(3)
```
      2 0
  3)6 2
    6
      2
      0  } You can omit it.
      2
```

STEP 35

(P.82 · 83)

■ Review of Step 34
(1) 26 R 1   (3) 29 R 2
(2) 19

1 (1) 12   (2) 11 R 1

2 (1) 16 R 1   (3) 18 R 1
(2) 17 R 1

3 (1) 13 R 2   (3) 15 R 3
(2) 15

© Kumon Publishing Co., Ltd.   137

4 (1) 12 R 2   (3) 16 R 3
    (2) 13 R 3

STEP 36

(P.84 · 85)

■ Review of Step 35

  (1) 16 R 3   (3) 14
  (2) 15 R 4

1 (1) 11 R 2   (2) 10 R 3

2 (1) 10 R 6   (3) 12
    (2) 12 R 2

3 (1) 10 R 1   (3) 10 R 7
    (2) 10

4 (1) 10   (3) 11
    (2) 10 R 5

STEP 37

(P.86 · 87)

■ Review of Step 36

  (1) 12 R 1   (3) 10 R 8
  (2) 12

1 (1) 321   (3) 243 R 1
    (2) 212

2 (1)
```
      2 1 0
   2) 4 2 1
      4
        2
        2
          1
```
    (2) 320 R 2
    (3) 220 R 3
    (4) 320
    (5) 120

3 (1)
```
      2 0 8
   3) 6 2 5
      6
        2 5
        2 4
            1
```
    (2) 208 R 2
    (3) 105 R 1
    (4) 403 R 1
    (5) 109 R 4

STEP 38

(P.88 · 89)

■ Review of Step 37

  (1) 211 R 2   (3) 102 R 4
  (2) 112 R 3

1 (1) 178   (2) 227

2 (1) 178 R 1   (6) 184 R 2
    (2) 286 R 1   (7) 226 R 3
    (3) 155 R 2   (8) 124 R 4
    (4) 195 R 1   (9) 120 R 6
    (5) 135 R 2

STEP 39

(P.90 · 91)

■ Review of Step 38

  (1) 169 R 1   (3) 117
  (2) 126

1 (1)
```
        8 5
   5) 4 2 6
      4 0
        2 6
        2 5
            1
```
    (2) 84 R 1
    (3) 86 R 1

2 (1) 64 R 1   (4) 31 R 2
    (2) 92   (5) 91 R 1
    (3) 72 R 2

3 (1) 60 R 3   (3) 60 R 5
    (2) 80

TEST

(P.92 · 93)

Review of Step 33-36

  (1) 48   (4) 11 R 1
  (2) 11 R 1   (5) 11 R 5
  (3) 13 R 2

138   © *Kumon Publishing Co., Ltd.*

Review of Step 37

(1) 311 R 1
(2) 228 R 1
(3) 105 R 1

(4) 322 R 1
(5) 221
(6) 340

Review of Step 38

(1) 139 R 1
(2) 123 R 7
(3) 146 R 4

(4) 226 R 2
(5) 124 R 2

Review of Step 39

(1) 57
(2) 73
(3) 77

(4) 52
(5) 40
(6) 81 R 2

STEP 40 (P.94・95)

■ Review of Step 39

(1) 93
(2) 81

(3) 70 R 3

❶ (1)

$$
\begin{array}{r}
1 \\
21{\overline{\smash{\big)}\,23}} \\
\underline{21} \\
2
\end{array}
$$

(2) 2 R 2
(3) 1 R 5
(4) 2 R 1
(5) 3 R 3
(6) 4 R 1

❷ (1)

$$
\begin{array}{r}
1 \\
21{\overline{\smash{\big)}\,35}} \\
\underline{21} \\
14
\end{array}
$$

(2) 2 R 3
(3) 2 R 13
(4) 3 R 2
(5) 3 R 12
(6) 4 R 8

❸ (1) 2 R 5
(2) 2 R 14
(3) 2 R 23

(4) 2 R 7
(5) 2 R 12

STEP 41 (P.96・97)

■ Review of Step 40

(1) 2 R 2
(2) 2 R 3

(3) 2 R 4

❶ (1) 1 R 2
(2) 1 R 14
(3) 1 R 27

(4) 2 R 13
(5) 2 R 22
(6) 3 R 2

❷ (1) 1 R 2
(2) 1 R 14
(3) 1 R 21

(4) 1 R 30
(5) 2 R 3

❸ (1) 2 R 1
(2) 3
(3) 3 R 2

(4) 2 R 4
(5) 2 R 1
(6) 2 R 9

STEP 42 (P.98・99)

■ Review of Step 41

(1) 1 R 3
(2) 2 R 2

(3) 1 R 16

❶ (1) 4 R 4
(2) 4
(3) 3 R 19

(4) 2 R 1
(5) 1 R 20

❷ (1) 3 R 7
(2) 3 R 17
(3) 4 R 6
(4) 2 R 3

(5) 2
(6) 1 R 29
(7) 3 R 1
(8) 2 R 18

❸ (1) 3
(2) 2 R 21

(3) 3 R 20

STEP 43 (P.100 · 101)

■Review of Step 42

(1) 2 R 12 (3) 3 R 18
(2) 2 R 19

❶ (1) 2 R 1 (4) 2 R 2
(2) 1 R 22 (5) 1 R 31
(3) 2 R 21 (6) 2 R 29

❷ (1) 1 R 20 (8) 2
(2) 3 (9) 1 R 29
(3) 2 R 21 (10) 2 R 1
(4) 2 R 1 (11) 1 R 31
(5) 1 R 42 (12) 2 R 15
(6) 1 R 37 (13) 3
(7) 2 R 2 (14) 2 R 27

STEP 44 (P.102 · 103)

■Review of Step 43

(1) 1 R 21 (3) 2 R 10
(2) 1 R 31

❶ (1) 3 R 1 (4) 2 R 15
(2) 1 R 23 (5) 2 R 25
(3) 1 R 29

❷ (1) 2 R 6 (8) 3 R 23
(2) 2 R 15 (9) 2 R 21
(3) 3 R 11 (10) 2 R 16
(4) 2 R 24 (11) 2 R 28
(5) 1 R 33 (12) 2 R 6
(6) 1 R 43 (13) 2 R 1
(7) 1 R 43 (14) 1 R 35

STEP 45 (P.104 · 105)

■Review of Step 44

(1) 3 R 15 (3) 3 R 20
(2) 3 R 18

❶ (1) 1 R 38 (4) 1 R 26
(2) 2 R 12 (5) 3 R 3
(3) 2 R 19

❷ (1) 2 R 22 (8) 2 R 10
(2) 1 R 33 (9) 3 R 2
(3) 3 R 12 (10) 3 R 8
(4) 2 R 20 (11) 1 R 42
(5) 2 R 15 (12) 1 R 30
(6) 3 R 17 (13) 2 R 23
(7) 2 R 20 (14) 2 R 20

STEP 46 (P.106 · 107)

■Review of Step 45

(1) 7 R 1 (3) 1 R 37
(2) 2 R 17

❶ (1) 5 R 11 (4) 2 R 25
(2) 7 R 10 (5) 2 R 26
(3) 4 R 10

❷ (1) 3 R 12 (9) 4 R 4
(2) 5 R 8 (10) 4 R 1
(3) 6 R 2 (11) 3 R 5
(4) 4 R 9 (12) 2 R 26
(5) 5 R 1 (13) 2 R 10
(6) 5 R 4 (14) 3 R 4
(7) 3 R 6 (15) 2 R 25
(8) 3 R 9

Review of Step 40

(1) 1 R 3 (4) 1 R 15
(2) 3 R 5 (5) 4
(3) 2 R 3

Review of Step 41,42

(1) 3 R 3 (4) 2 R 10
(2) 2 R 5 (5) 1 R 16
(3) 1 R 3 (6) 2 R 5

Review of Step 43-45

(1) 2 R 9 (4) 2 R 28
(2) 3 R 20 (5) 2 R 31
(3) 5

Review of Step 46

(1)
```
      7
12 ) 93
     84
      9
```

(4)
```
      5
15 ) 76
     75
      1
```

(2)
```
      4
13 ) 64
     52
     12
```

(5)
```
      4
16 ) 67
     64
      3
```

(3)
```
      3
14 ) 55
     42
     13
```

(6)
```
      3
18 ) 55
     54
      1
```

■ Review of Step 46

(1) 7 R 8 (2) 3 R 2

❶ (1) 5 R 4 (3) 6 R 3
(2) 5 R 14 (4) 7 R 1

❷ (1) 4 R 1 (3) 4 R 4
(2) 6 R 1 (4) 6 R 2

❸ (1) 2 R 16 (4) 6 R 22
(2) 3 R 10 (5) 5 R 34
(3) 4 R 10 (6) 3 R 15

■ Review of Step 47

(1) 6 R 1 (2) 5 R 21

❶ (1) 4 R 6 (3) 7 R 2
(2) 6 (4) 8 R 6

❷ (1) 7 R 2 (6) 9
(2) 8 R 6 (7) 9 R 20
(3) 8 R 10 (8) 6 R 12
(4) 8 R 4 (9) 5 R 2
(5) 8 R 26 (10) 5 R 7

■ Review of Step 48

(1) 8 R 3 (2) 4 R 20

❶ (1) 5 R 23 (3) 7 R 49
(2) 6 R 9 (4) 8 R 3

❷ (1) 7 R 13 (6) 7 R 55
(2) 8 R 40 (7) 7 R 58
(3) 7 R 51 (8) 8 R 38
(4) 8 R 20 (9) 6 R 92
(5) 8 R 70 (10) 7 R 68

■ Review of Step 49

(1) 8 R 30 (2) 7 R 4

❶ (1) 11 R 14 (2) 21 R 4

② (1) 2| R 23   (4) 23 R 34
(2) 2| R 2| (5) 42 R 3
(3) 2| R || (6) 27 R 10

STEP 51
(P.118 · 119)

■ Review of Step 50

(1) 2| R 23   (2) 35 R ||

① (1)
```
        23
  38)895
     76
    135
    114
     21
```
(2)
```
        26
  27)713
     54
    173
    162
     11
```

② (1)
```
        24
  32)769
     64
    129
    128
      1
```
(5)
```
        26
  35)941
     70
    241
    210
     31
```
(2)
```
        36
  27)975
     81
    165
    162
      3
```
(6)
```
        26
  38)998
     76
    238
    228
     10
```
(3)
```
        36
  26)941
     78
    161
    156
      5
```
(7)
```
        37
  23)864
     69
    174
    161
     13
```
(4)
```
        36
  26)936
     78
    156
    156
      0
```
(8)
```
        28
  34)974
     68
    294
    272
     22
```

STEP 52
(P.120 · 121)

■ Review of Step 51

(1) 2 R 20   (2) 34

①
```
        20
  32)654
     64
     14
```

② (1)
```
        40
  14)572
     56
     12
```
(5)
```
        30
  26)784
     78
      4
```
(2)
```
        30
  24)724
     72
      4
```
(6)
```
        30
  29)897
     87
     27
```
(3)
```
        60
  14)852
     84
     12
```
(7)
```
        30
  32)973
     96
     13
```
(4)
```
        40
  24)975
     96
     15
```
(8)
```
        20
  46)926
     92
      6
```

STEP 53
(P.122 · 123)

■ Review of Step 52

(1) 30 R 12   (2) 30 R |

① (1) 6 R 13   (3) 3 R 39
(2) 4

② (1) 40 R 56   (4) 3| R 203
(2) 35 R 12| (5) 30 R 186
(3) 32 R 240

■ Review of Step 53

(1) 2 R 135   (2) 34 R 34

❶   323 R 6

❷ (1)
```
      270
  21)5678
     42
    ───
    147
    147
    ───
       8
```

(4)
```
      200
  32)6421
     64
    ───
      21
```

(2)
```
      287
  31)8902
     62
    ───
    270
    248
    ───
    222
    217
    ───
       5
```

(5)
```
      125
  32)4000
     32
    ───
      80
      64
    ───
     160
     160
    ───
       0
```

(3)
```
      219
  31)6799
     62
    ───
     59
     31
    ───
    289
    279
    ───
     10
```

(6)
```
      161
  42)6789
     42
    ───
    258
    252
    ───
     69
     42
    ───
     27
```

TEST   (P.126 · 127)

Review of Step 47,48

(1) 7 R 2   (3) 5 R 2

(2) 8   (4) 5 R 3

Review of Step 49,50

(1) 7 R 30   (3) 12 R 8

(2) 7 R 33   (4) 21 R 13

Review of Step 51,52

(1) 22 R 31   (3) 30 R 25

(2) 25 R 17   (4) 30 R 1

Review of Step 53,54

(1)
```
        2
  342)719
      684
     ───
       35
```

(3)
```
       165
  41)6789
     41
    ───
    268
    246
    ───
    229
    205
    ───
     24
```

(2)
```
         22
  362)7986
      724
     ───
      746
      724
     ───
       22
```

(4)
```
       202
  44)8902
     88
    ───
    102
     88
    ───
     14
```